Daniel Giraud Elliot

# North American Shore Birds

A history of the snipes, sandpipers, plovers and their allies, inhabiting the beaches and marshes of the Atlantic and Pacific coasts, the prairies and the shores of the inland lakes and rivers of the North American continent. Second Edition

Daniel Giraud Elliot

**North American Shore Birds**
*A history of the snipes, sandpipers, plovers and their allies, inhabiting the beaches and marshes of the Atlantic and Pacific coasts, the prairies and the shores of the inland lakes and rivers of the North American continent. Second Edition*

ISBN/EAN: 9783337239510

Printed in Europe, USA, Canada, Australia, Japan

Cover: Foto ©berggeist007 / pixelio.de

# NORTH AMERICAN SHORE BIRDS

A HISTORY OF THE

## SNIPES, SANDPIPERS, PLOVERS AND THEIR ALLIES

INHABITING THE BEACHES AND MARSHES OF THE ATLANTIC AND PACIFIC COASTS, THE PRAIRIES, AND THE SHORES OF THE INLAND LAKES AND RIVERS OF THE NORTH AMERICAN CONTINENT;

THEIR POPULAR AND SCIENTIFIC NAMES, TOGETHER WITH A FULL DESCRIPTION OF THEIR MODE OF LIFE, NESTING, MIGRATION AND DISPERSIONS, WITH DESCRIPTIONS OF THE SUMMER AND WINTER PLUMAGES OF ADULTS AND YOUNG, SO THAT EACH SPECIES MAY BE READILY IDENTIFIED.

*A Reference Book for the Naturalist, Sportsman and Lover of Birds*

BY

DANIEL GIRAUD ELLIOT, F.R.S.E., ETC.

EX-PRESIDENT AMERICAN ORNITHOLOGISTS' UNION

*Curator of Zoölogy in the Field Columbian Museum, Chicago; Author of "Birds of North America," Illustrated Monographs of Ant Thrushes, Grouse, Pheasants, Birds of Paradise, Hornbills, Cats, Etc.*

WITH SEVENTY-FOUR PLATES

*SECOND EDITION*

NEW YORK
FRANCIS P. HARPER
1895

## PREFACE TO THE SECOND EDITION.

THE favorable reception accorded to the SHORE BIRDS, a large edition having been exhausted in a comparatively brief period, has proved that, although the number of works devoted to ornithology, which are being continually issued, is very large, there was yet room for a popular one such as this. The author is gratified to know that, in a very great degree, his book has reached the particular classes—sportsmen and students of birds and bird life—for which it was especially written, and trusts that his efforts to make them better acquainted with these fairy wanderers of our land have met with some degree of success.

In preparing this edition, the letterpress has been carefully examined and the few typographical errors that may have existed in the first edition have been corrected. The kindly criticisms, also, of my colleagues on these matters have been of considerable assistance. In the Appendix the Key to the Families has been slightly rearranged, but not changing in any way the definitions.

All species that had any claim to be considered North American were intended to be included in the first edition, and indeed the door of admission was thrown so widely open that possibly in a rare instance a species stepped in which might, without committing an injustice, be regarded somewhat in the light of an intruder, as it is not probable that any of my readers would ever meet with it in the flesh. But it seems there are yet claimants to the honor of American citizenship, and although it is called *primaries*, the ornithological term, because they are the first or most important, and without which no bird could rise and sustain itself in the air. They are sometimes called "flight feathers," but this term carries with it no clearer definition to the layman than "primaries," and consequently attains the desired result of simplicity or clearness no better.

The long sentence given above is, of course, impossible, objectionable in two ways—the space it occupies and the weariness that would arise from its constant repetition. In order, therefore, to render clear any term referring to, or describing any portion of, a bird's plumage, a "map" of a bird is given, and every part indicated, with the name it bears, clearly portrayed, thus serving as a handy dictionary, and explaining all the terms used in these pages. It is believed that with little trouble the use of this "map" will smooth away any difficulty arising from the occurrence of some unfamiliar word in the descriptions. A glossary will also be found defining all the terms used.

To the author, the "Shore Birds" have been, from his boyhood, objects of special attraction, both for the sport they afforded and for their elegant forms and gentle natures; and it has been a great source of pleasure to watch them in their habitats whenever an opportunity occurred. From far Alaska and the Pacific coasts of the United States, across its wide domain in nearly every State of the Union; and from the Gulf of St. Lawrence on the east to the Gulf of Mexico in the south, and along the Atlantic coasts, over to the adjacent islands such as Bermuda, Cuba, and the Windward Islands to Rio Janeiro in South America, he has shot and observed these little wanderers, and the major part of the accounts of their habits is derived from his own experience.

The nesting habits of certain species which resort within the Arctic Circle for the purposes of incubation, have been observed only by those who have penetrated those far Northern districts, and the accounts of these have been to a great extent derived from MacFarlane, Nelson, Murdoch, Dall, and Bannister, whose experience in the Barren Grounds and Alaskan Peninsula and along the shores, and among the islands of Behring Sea, has enabled them to place on record valuable and interesting facts relating to the birds met with. A large proportion of the species mentioned in these pages breed within our own limits, and their ways and manners during that interesting period can be observed by any one who takes an interest in the subject.

The tables at the end of the volume are constructed on the simplest plan possible, showing by graduated steps the various characters possessed by each species, as well as those distinguishing one bird from another. The one first given is an attempt to indicate the differences that exist between the families; this is followed by a table defining the genera, and under each genus as it is reached is a table which characterizes every species. So, if a Shore Bird is procured unknown to its possessor, he must first look in the table for Families to see if it should be classed among Phalaropes, Stilts, Snipe, Plover, etc., and then to the table of genera in the family, to ascertain in which it is included. If the bird is not at once determined, the probability is that the investigator will be brought so near the right species as to find it in a short time. In a large group of birds, whose members have so many points of resemblance, it is not to be expected that any scheme can be devised which will enable one without knowledge of the subject immediately to determine any bird he may observe or procure, but by a little patience and slight familiarity in the use of these tables, the desired result can surely be obtained.

The Latin names given at the head of the descriptions are, with one or two exceptions, those employed in the Check List of the American Ornithologists' Union.

In measuring a bird so as to compare it with the various dimensions given for each species, *total length* means from the tip of the bill to end of tail as the specimen lies extended on the table; *wing*, from the bend or shoulder to end of longest primary; *tail*, from the lump just below the rump to end of longest feather; *bill* (when *culmen* is stated), from the beginning of the feathers on the forehead along the ridge to the extreme point; *tarsus*, from the joint or heel to root of toes; *middle toe*, from its connection with the tarsus to beginning of the claw. A scale drawn to one-tenth of an inch will be found on a separate page, affording a handy measure for obtaining dimensions of specimens.

The birds from which the descriptions have been taken, also of a large proportion of the eggs, are in the collections of the American Museum of Natural History in New York, and in those of my friend, Mr. George B. Sennett, deposited in the same institution.

The illustrations which ornament the volume and add so greatly to its usefulness and attractiveness are drawn by Mr. Edwin Sheppard, of the Academy of Natural Sciences of Philadelphia, an artist possessing exceptional talent for portraying birds and bird life.

# CONTENTS.

# LIST OF ILLUSTRATIONS.

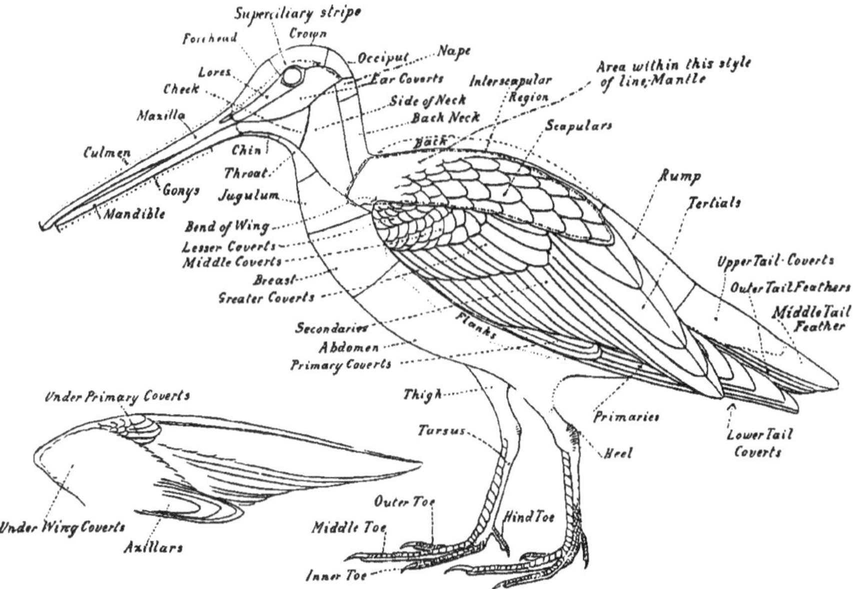
Superciliary stripe
Crown
Forehead
Occiput
Nape
Lores
Ear Coverts
Cheek
Interscapular Region
Area within this style of line, Mantle
Side of Neck
Maxilla
Back Neck
Scapulars
Culmen
Chin
Back
Throat
Rump
Gonys
Jugulum
Tertials
Mandible
Bend of Wing
Lesser Coverts
Middle Coverts
Upper Tail-Coverts
Outer Tail Feathers
Breast
Middle Tail Feather
Greater Coverts
Flanks
Secondaries
Abdomen
Primary Coverts
Under Primary Coverts
Thigh
Primaries
Lower Tail Coverts
Tarsus
Heel
Outer Toe
Hind Toe
Middle Toe
Under Wing Coverts
Axillars
Inner Toe

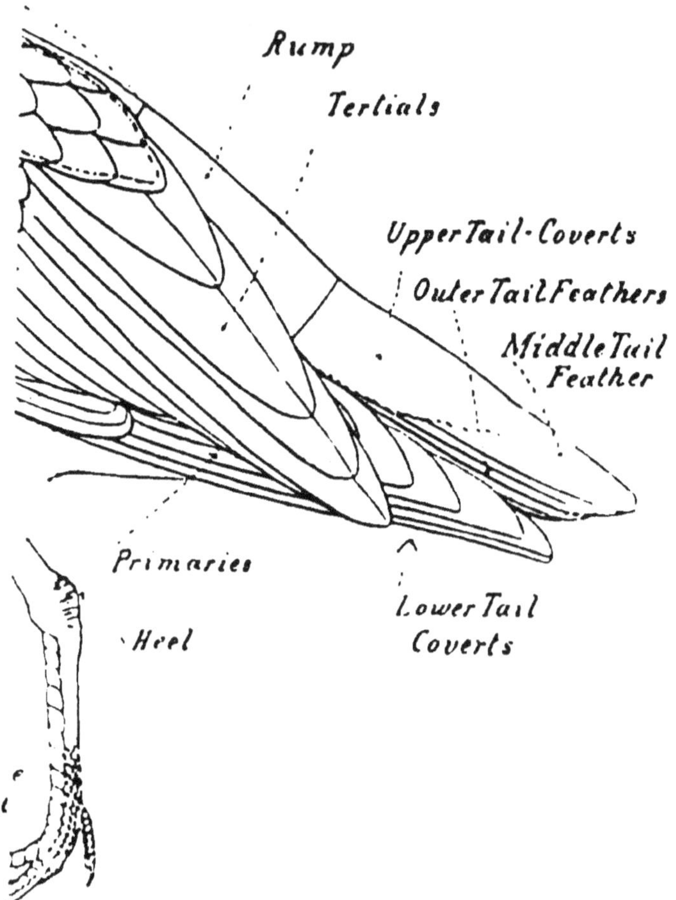
Area within this style
of line, Mantle
Scapulars
Rump
Tertials
Upper Tail-Coverts
Outer Tail Feathers
Middle Tail
Feather
Primaries
Heel
Lower Tail
Coverts

# GLOSSARY.

MAXILLA. The upper half of bill.

CULMEN. The middle lengthwise line of maxilla.

MANDIBLE. The lower half of bill, usually slightly shorter than the maxilla.

GONYS. Lower outline of mandible from tip to point of separation.

GAPE. Cleft between maxilla and mandible when open.

LORES. Space between eye and base of maxilla.

FOREHEAD. Top of head from base of bill to crown.

SUPERCILIARY STRIPE, OR SUPERCILIARIES. Line running usually from base of maxilla over the eye to side of occiput.

CROWN. Space on top of head from front line of eye to beginning of back slope of skull.

OCCIPUT. Hind part of skull, forming back slope of head.

NAPE. Upper portion of back of neck.

EAR-COVERTS. Loose-webbed small feathers overlying the ears.

AURICULAR REGION. Space about external opening of the ears.

CHEEK. Space between angle of jaw and bill.

SIDE OF NECK. Lateral portions of neck.

BACK OF NECK. Portion from occiput to back.

BACK. Space lying between base of hind-neck and rump.

SCAPULARS. Feathers alongside the back, and overlying inner line of wing-coverts.

INTERSCAPULAR REGION. Feathers of back lying between the scapulars

MANTLE. Back, scapulars and upper surface of wing.

BEND OF WING. Angle of wrist-joint forming the foremost end of wing when closed.

LESSER COVERTS. Several rows of small feathers near bend of wing.

MIDDLE COVERTS. Feathers lying below the lesser coverts, and which overlap each other in a reverse manner from the other coverts.

GREATER COVERTS. Rather large feathers lying below the middle coverts and overlying the base of the secondaries.

SECONDARIES. Feathers of varying length which are seated on the fore-arm (*ulna*), and sometimes sufficiently long to entirely cover the primaries.

TERTIALS. Inner feathers of the wing springing from the humerus, more or less covered by the longer scapulars.

PRIMARIES. Long feathers of the wing growing from the hand and finger bones.

PRIMARY COVERTS. A series of stiff feathers overlying the base of the primaries, frequently concealed beneath the secondaries.

UNDER PRIMARY COVERTS. Feathers covering the bases of the primaries on the under side of the wing.

UNDER WING COVERTS. Feathers covering the under side of the wing, forming a lining.

AXILLARS. Elongated graduated feathers growing from the arm pit, (*axilla*), and lying close to the body.

RUMP. Portion of the body between the back and upper tail-coverts.

TAIL. Stiff feathers of various lengths, born upon the last caudal vertebra, and forming a rudder by which a bird steers its course during flight.

MIDDLE TAIL FEATHERS. Broad feathers in the center which overlie and partly conceal the rest when the tail is closed.

UPPER TAIL-COVERTS. Feathers covering the base of the tail.

LOWER TAIL-COVERTS. Feathers beneath the tail.

CHIN. Space between the diverging branches of the mandible.

THROAT. Upper part of neck below the chin.

JUGULUM. Lower part of fore-neck between the throat and breast.

BREAST. Space between the jugulum and abdomen.

ABDOMEN. The belly.

FLANKS. Feathers on side of body, lying along the central portion of the wing.

THIGH. Bone between knee and heel.

HEEL. Upper extremity of tarsus.

TARSUS. Bone from heel to base of foot.

OUTER TOE. Toe on outside of foot in front, usually next in length to middle toe.

MIDDLE TOE. The central, generally longest toe of the foot in front.

INNER TOE. Smallest of front toes on inside of foot.

HALLUX, OR HIND TOE. Small toe on back of tarsus, usually above the level of front toes.

# INTRODUCTION.

## SNIPES, SANDPIPERS, PLOVERS, ETC.

THE great bird groups comprising the Snipes, Sandpipers, Plovers, and their allies form what is called the order *Limicolæ*—Shore Birds, literally—"living in the mud." They have considerable affinity to Gulls, also to Cranes and Bustards. One family, containing the Jacanas (singular birds with excessively lengthened toes, fitting them for their semi-aquatic life), while related to the Plovers in general structure, have also many affinities to the Rails. The three divisions into which the *Limicolæ* are divided—the Snipes. Sandpipers, and Plovers—differ from each other in structure, but not very greatly in habits. The bill in the majority of instances serves to distinguish the members of the first two divisions from the last, this in the Snipe and Sandpiper being rather long as a rule, frequently flexible and sensitive at the tip, with the ridge of the maxilla usually level; while the bill of the Plover is more pigeon-shaped, with the ridge of the maxilla (or upper portion of the bill) often considerably depressed near the head, and with a convex horny or hard portion at the tip, this part not flexible nor particularly sensitive. Absence of the hind toe was once deemed a strong, almost unfailing Plover character, and while it is still one used to indicate a Plover, yet it must not by any means be relied upon as sufficient, for some, even true Plovers, have at least the hind toe in a rudimentary form, and certain species of the other divisions are without any vestige of a hind toe. Snipes and Sandpipers have the feathers of the head stop abruptly at the base of the bill; the nostrils are placed in grooves which sometimes run nearly to the tip of the bill, and are narrow exposed slits. The body is usually slender, and the legs long, covered with transverse scales before and behind, and reticulated on the sides, with the exception of the Wandering Tattler and the Curlews, which have the hind part of the leg covered with small hexagonal scales. In the North American species the hind toe is absent only in the Stilt and Sanderling, all the other species having this member well developed. Sometimes the toes are webbed at the base; either all the anterior toes are so furnished, or a web is present between the middle and one of the lateral toes. In some cases also lateral membranes are present on the toes, even or scalloped. The Snipes are apt to obtain their food by probing for it in the mud or sand, the sensitive tip indicating the presence and quality of the object touched by it. In the Plovers the feathers of the forehead do not cease so abruptly as in the Snipes, but in some instances reach the nostrils. These are short, rather wide, and open in a skin which fills the nasal groove. The eye is very large and luminous, and the head is large, rounded, sloping in front rapidly to the bill. The hind toe is usually absent, though in one species of true Plover (Black-breasted) it is present in a rudimentary form, and is well developed in the Turnstones and entirely wanting in the Oyster-catchers. In Jacana, of course, it is obtrusively present, with a claw exceeding the toe in length. The legs are usually rather short and covered with small scales, but some, as the Turnstones, have the forepart covered with transverse scales. The body is fuller, shorter, and more rounded than is that of the Snipe, and the general appearance is not so graceful and elegant.

All the members of the order have long pointed wings and rather short tails. The primaries are stiff and narrow, the first being the longest and graduating rapidly to the last or innermost, the difference of length between these extremes affording an excellent character for one genus of Snipe—*Tringa*. The secondaries rapidly lengthen in the reverse order, the innermost frequently reaching to the tail, forming between these two classes of feathers, when the wing is spread, a deep emargination. The young of this order are what is called *Præcoces*, or those that are able to run about as soon as they emerge from the shell, in contradistinction to the *Altrices*, or those that remain in the nest until they are able to fly. They are covered with a soft down, which in its distribution of colors indicates somewhat the markings of the adult dress, but which, from its close similarity to the various hues and tints of the surrounding locality in which the nest is placed, renders detection of their presence almost impossible. In some species, mainly the Woodcock and true Snipe, the adult dress does not change in its general appearance throughout the year, but with the majority, the summer or breeding dress is succeeded in the autumn by a winter plumage, of which the principal hues are black, white, and gray. Only in occasional instances do any brilliant and conspicuous colors appear, but as a rule the combinations of various tints and the distribution of even strongly contrasting hues are so beautifully arranged and harmoniously adjusted that all the species present a most attractive appearance and constitute very agreeable features in the landscape.

Snipes and Plovers moult twice in the year—spring and autumn—and the first plumage of the young resembles somewhat the adult summer dress; but this soon changes to the winter plumage of their parents, and is the garb in which they present themselves, during their first migration, to the inhabitants of lands south of their breeding places. In some instances, from use and exposure, the feathers of the body become shortened by the wearing away of their margins, and then the bird presents a different aspect, as though it had almost moulted, the brighter colors of the summer dress having given way to the more subdued tints prominent in autumn, as the body colors or those lying towards the center of the feather come into view. These birds, properly speaking, make no nest, a slight depression in the ground, sometimes lined with grass, sufficing for the purpose, and the eggs, generally pyriform in shape, are dotted and spotted with various colors. The order is a very large one, represented in all parts of the globe, and contains about two hundred species.

2. Red Phalarope.

## RED PHALAROPE.

THE Red or Gray Phalarope, Coot-footed Tringa, Sea-goose, Whale Bird, and Bowhead Bird, by each and all of which names it is known, is, like its relatives, a bird of the boreal regions, coming southward only when driven by the severity of the winter's cold, when it appears along our coasts and in the Ohio Valley. But it is more a maritime than an inland species, and finds its home upon the waves, being frequently seen one to two hundred miles from land. It has been procured on Long Island, though rather rare, and I had a fine specimen in my possession in winter plumage, killed near Islip. It is doubtful if it goes on our Atlantic coast farther south than New Jersey, while in the autumn it is occasionally seen on the Western lakes and rivers, and on the Pacific Coast has been obtained as far south as Cape St. Lucas. It keeps in flocks, swims like a duck upon the water, and when at sea frequents the beds of floating weed and grass, on which it runs as though on solid ground, and feeds upon the minute crustacea found clinging to the leaves. In northern latitudes it goes in large flocks, and is found in numbers in Alaska at Point Barrow, and at the mouth of the Yukon, in May and June. In the latter month they are usually mated, and are scattered over the grassy flats in couples feeding, and are tame and unsuspicious. Hunger satisfied, a few will rise on the wing and fly over the flats. These are soon joined by others, until at times several hundred may have gathered together, pursuing a most erratic course, as they rapidly pass over the land ; at one time rising high in the air, executing swift and graceful evolutions, then descending and skimming along just over the ground, moving in perfect unison and twisting from one side to the other with such regularity as to show alternately the upper and under side of the body, with its intermittent flashes of light and shade. Wearied at length, they again seek the grassy flats and scatter in quest of food.

The female is the larger and handsomer bird, and she does the courting, while the male performs most of the duties of incubation, thus affording an instance of the exercise of "woman's rights" in the fullest degree. The nest is merely a depression in the damp flats, usually without any lining, in generally pretty wet situations, and the number of eggs is mostly four. They vary greatly in color, some are pale buffy brown tinged with olive, others pale greenish gray, or sea green, profusely blotched and spotted with purplish or dark brown, with underlying spots of pale grayish brown, largest, and running together frequently, on the larger end. They vary in size from .85 × 1.07 inches to .88 × 1.27. This species breeds all along the Arctic shores of Alaska and Siberia, also in Greenland, Iceland, and Spitzbergen, and in the last island the eggs have been found laid upon the bare ground. As soon as the young are able to fly, the birds gather in flocks and pass to the sea, though occasionally they frequent the ponds back of the beach. By the beginning of August they have commenced to assume the gray winter plumage, and the old birds start on their southern migration, those that remain until late in the autumn being apparently the young of the year. The names "Whale Bird" and "Bowhead" are given to this Phalarope by the whalers, for the reason that they feed upon the animalculæ which form the food of the right whale or bowhead, and so it follows that whenever a large number of these elegant birds are found congregated together at sea, the presence of whales is predicted with considerable certainty. The dreary expanse of ice-dotted sea is frequently enlivened by the graceful forms of this beautiful species, which find sheltered resorts in the calm open spaces amid the great frozen field. They are among the last of their tribe to leave the cheerless North for summer climes.

## *CRYMOPHILUS FULICARIUS.*

*Habitat.*—North portions of Old and New Worlds, in winter going south on Atlantic Coast to Middle States, and to Chili, on the Pacific. Ohio Valley—mostly a maritime species. Breeding in Arctic regions from Greenland to Alaska.

*Adult Female in Summer.*—Nape, crown, forehead, sides of bill and chin, black; sides of head and line around the eye, and another below the nape, pure white; neck (except a narrow black and plumbeous line on the hind part) and entire under parts, deep purplish cinnamon; back and scapulars, black, feathers margined with buff; primaries, brownish on outer web, blackish at tip, and brown, graduating into silvery white, on the inner web; secondaries, brown, margined with white, one or two short ones almost pure white; long inner secondaries, black margined with buff, or buffy white; wing-coverts, dark gray, the median ones edged with whitish, the greater ones margined with white, forming a bar across the wing; rump, plumbeous in center, white on sides; upper tail-coverts, cinnamon or rusty red, the center of some of the middle ones black; middle tail-feathers, black, remainder slate gray, the two outer ones dark rufous at tip; bill, flat, yellow, or orange yellow, black at tip; legs and feet, dull olive; eyes, brown. Length, 7½–8 inches; wing, 5¼–5½; tail, 2½; tarsus, ⅞; culmen, 80–95.

*Male in Summer.*—Similar to female, but duller in color. Feathers on crown and nape marked with rusty or yellowish brown. The white on sides of head is smaller and ill-defined, and feathers on abdomen are edged with white.

*Winter Plumage.*—Occiput and around eyes, with a narrow line on hind-neck and upper part of back, black; rest of head, neck, and entire under parts, pure white; back and scapulars, dark gray; wings and rump, brownish black; wing-coverts and secondaries, edged with white; tail, brownish black, outer feathers palest; bill, dark, almost black; legs, grayish olive.

*Young, First Plumage.*—"Crown, nape, back, and scapulars, dull black, the feathers edged with ochraceous; wing-coverts, rump, and upper tail-coverts, plumbeous, the middle coverts bordered with pale buff, the tail-coverts with ochraceous; head (except crown) and lower parts generally, white; the throat suffused with brownish buff."—*Ridgway.*

*Downy Young.*—"Above, bright tawny buff, marked with broad irregular stripes of black; superciliary stripes, bright tawny buff, separated only by a narrow and sometimes interrupted dusky streak; crown, bright raw-umber brown bordered with black; chin and throat, light fulvous buff, changing to smoky buff on chest; rest of lower parts, dull whitish."—*Ridgway.*

3. Northern Phalarope.

## NORTHERN PHALAROPE.

A CIRCUMPOLAR species, the Northern or Red-necked Phalarope is perhaps not so abundant in the extreme north as the Red Phalarope. It breeds on the islands in Behring Sea, as well as on the north coast of Siberia, and is common in the interior of Northern Alaska. It is a summer visitor to Greenland, Iceland, and the Faroes, and breeds on Nova Zembla and the New Siberian Islands, also above the pine regions of the Doorefjeld as far south as latitude 62°, and Middendorf found it breeding on the Pacific Coast on the west shore of Okhotsk Sea to south latitude 55°. It is a lovely, graceful species, gentle and unsuspicious, at home upon the bosom of the ocean, frequenting the floating masses of kelp and other weeds and grasses, seeking the small worms, insects, and minute crustaceans, which form its principal food. On the water this species floats with airy lightness, and moves as easily as a feather when wafted by a gentle breeze over the surface of a placid pool. It advances gracefully, keeping time with each stroke of the feet by a motion of its head and neck, turns rapidly and snatches some microscopic morsel from the surface of the water. It goes in flocks, constantly uttering a sharp metallic *tweet* or *twick*, and is always in motion, swimming over the surface or flying from one mass of weed to another. Early in May it arrives at the breeding grounds, and the females commence to make love to the apparently indifferent males, using all the wiles and blandishments generally employed by one of the sterner sex (when bent on a like purpose) to gain the favor and secure the affection of the object of its adoration. And the male is as coy and retiring as the most bashful maiden, turning away from the proffered attentions, first to this side, then to that, even flying to the opposite side of the pool, or to another near by; but all in vain, for he is followed by the fair one who has chosen him from his fellows, and there is no escape. He swims rapidly along, but she is ever near, and with arched neck circles about him, rising on wing at times and poising above him, and producing a sharp series of sounds by quick strokes of the pinions. At last, like any other poor bachelor so beset, he yields, and the nest, a slight structure of dry stalks, is placed in the center of a thick tuft of grass. The eggs are four in number, pale, or rich buff, or pale olive, blotched and spotted with blackish brown and pale brown, with underlying grayish markings. On these the poor male, a victim to woman's rights, is obliged to sit the greater part of the time, the female amusing herself on the pool near by. If disturbed from the nest, the birds will fly to the water and swim about apparently unconcerned, moving their heads and necks with quick jerks and sipping the water with their bills. By the middle of July the young are able to fly, and towards the end of the month a few of the autumnal feathers commence to appear, and old and young gather together in flocks of a hundred or more and assemble on the shores, near the borders of ponds or rivers, where they remain until the last of September, when they seek milder climes. The flight of this species is at times very rapid, and the bird frequently twists and turns in a zigzag course like the common Snipe. I have frequently seen large flocks of this beautiful species in the "inland passage" of Alaska, and about the mouths of the rivers, disporting themselves on the water, or flying above its surface in many graceful evolutions. It was always very tame and gentle, allowed one to approach quite near, and constantly uttered a low tweet, as though the different individuals were carrying on a general conversation. It is a very sociable bird, and large numbers live together apparently in most amiable intercourse. In winter the Northern Phalarope goes frequently far to the south, and has been taken in the Bermudas, Guatemala, and the Isthmus of Tehuantepec.

## *PHALAROPUS LOBATUS.*

*Habitat.*—Arctic regions of both hemispheres, where it breeds to north latitude 73° in the western, and to limits of forest in the eastern, section. South in winter to the Tropics.

*Adult Female in Summer.*—Head, hind-neck, and back, dark plumbeous, the feathers on back and scapulars margined with rusty buff; a white spot above and below the eye; sides and front of neck and upper part of breast, chestnut, bordered beneath on the breast by a line of plumbeous, narrowest in the center; chin and upper part of throat and rest of under parts, pure white; flanks, white, streaked with sooty black; wings, blackish brown, greater coverts margined with white, forming bar across wing; primaries, brownish black; center of rump, black, sides white; tail, blackish brown; bill, black; legs and feet, lead color; iris, dark brown. Total length, 7 inches; wing, 4; tail, 2; tarsus, .75–.85 inch.

*The Male* is smaller and of not so bright a plumage; the head is sooty, marked slightly with light brown, and the back is more marked with ochraceous. The sides of the neck and breast are not so deep a chestnut and not so sharply defined. Over the ears there is a trace of the white eye-stripe.

*Adult in Winter Plumage.*—Forehead, stripe over the eye, chin, cheeks, throat, and rest of under parts, pure white; a black spot before the eye; crown, gray; a mixed black and gray stripe under the eye; sides of neck, white, washed with brownish yellow; hind-neck, bluish gray; back and wings, bluish gray, feathers margined with grayish white; rump and tail, blackish brown, the central tail-feathers bordered with grayish white.

*Young, First Plumage.*—Crown, plumbeous dusky, sometimes streaked; back and scapulars, blackish, margined with buff; innermost secondaries, upper tail-coverts, and tail, dark brown, margined with chestnut; forehead, front of the eye, and under parts, white; sides of breast, washed with brown.

*Downy Young.*—"Above, bright tawny, the rump with three parallel stripes of black, inclosing two of lighter fulvous than the ground color; crown covered by a triangular patch of mottled darker brown, bounded irregularly with blackish; a black line over ears not reaching to the eye; throat and rest of head, light tawny fulvous; rest of lower parts, white, becoming grayish posteriorly."—*Ridgway.*

4. Wilson's Phalarope.

## WILSON'S PHALAROPE.

ONE of the most beautiful of all our waders, this handsome and graceful bird is restricted to the New World, and is more of an inland species, rarely visiting the seacoast. It is abundant and generally distributed throughout the Mississippi Valley, and does not frequent very high northern latitudes like its relatives, but, on the other hand, penetrates much farther southward than any other species of Phalarope. On the eastern portion of the United States it is rather a rare bird, being occasionally met with on the seacoast from Massachusetts to New Jersey. It is quite common in Illinois, Iowa, Wisconsin, the Dakotas, Utah, and Oregon, in all of which States it breeds, and also on the Saskatchewan Plains, where Richardson found it breeding, but was not seen beyond the fifty-fifth parallel, nor at Hudson's Bay. As a rule, Wilson's Phalarope goes in small companies, though at times large flocks of several hundred are met with. It is not very shy, frequently permitting one to approach within a few feet, and it does not swim so much upon the water as is the habit of the other species, but wades about up to its belly picking its food from off the surface. When necessary, however, it swims gracefully and with ease, and the young soon after emerging from the egg are equally at home upon the surface of ponds, paddling about and diving with facility. The female is the larger and altogether the handsomer bird, the male having very little of the brilliant tints which render her so attractive when arrayed in her full summer dress. Upon him, too, devolves the duty of incubation to a very great degree, the female amusing herself upon or near the water. Like the other species of Phalarope, she makes all the advances at the pairing season, and sometimes more than one female fixes her affection upon some particular male, who thereupon has but little peace, as he is pursued from place to place by the rival suitors. Finally, the matter having been successfully arranged, the winged Dido bears off her Æneas, and a slight depression having been scratched in the soil and lined with grass, or a loosely constructed nest made in a clump of grass, the eggs, three or four in number, are deposited and the male assumes the novel and unusual duties for one of his sex, of incubation. The eggs vary from a fawn color to a rufous drab, profusely spotted and speckled with different shades of brown, thickest at the larger end, and measure .94 inch in breadth by 1.37 in length. Wilson's Phalarope is a rather silent species, its note having a kind of nasal quack-like sound. Its food is similar to that of the other Phalaropes.

### *STEGANOPUS TRICOLOR.*

*Habitat.*—Temperate North America, mainly inland. Breeds from Northern Illinois and Utah to the Saskatchewan region. In winter goes south to Patagonia.

*Adult Female in Summer.*—Forehead and top of head, pale bluish gray; nape and center of hind-neck, white; a narrow white line commencing half-way between nostril and eye, passing over the latter; line through the eye and auricular region, velvety black, graduating into rich deep chestnut, and continued as a narrow stripe on either side of the back to the tip of the scapulars; mantle, pearly gray; wings, dull pale brown, the greater coverts edged with white; primaries, dull brown; rump, dull brown; upper tail-coverts, dull brown margined with white; tail, pale brown, marked with white; fore-neck, cinnamon buff, graduating on the chest and flanks into white; chin, throat, cheeks, and under parts, pure white; bill, legs, and feet, black. Length, 9½–10 inches; wing, 5¼; tarsus, 1⅓; culmen, 1⅓.

*Adult Male in Summer.*—Smaller than the female and very much duller in color of plumage. Top of head, brown, feathers tipped with gray; broad mark over the eye and middle of hind-neck, white; sides of neck, dull rufous; back and wings, blackish, feathers margined with pale brown; rump and tail, brownish black, feathers edged with white; primaries, blackish brown; fore-neck, pale cinnamon; chin, cheeks, throat, and under parts, white. The brilliant markings of the female are either absent or faintly indicated, and he is a very plainly colored bird beside his beautiful mate. Length, 8¼–9 inches; wing, 4¾; culmen, 1¼; tarsus, 1¼.

*Winter Plumage.*—Upper parts, ash gray; rest of plumage, white, the breast shaded with pale gray.

*Young.*—Upper parts, blackish, feathers bordered with buff, distinct on inner secondaries; upper tail-coverts, superciliaries, and under parts, pure white, tinged with rusty on the breast; tail, ash, feathers edged and marbled with white.

*Downy Young.*—"Bright tawny, paler beneath, the belly nearly white; occiput and hind-neck, with a distinct median streak of black, on the former branching laterally into two narrow irregular lines; lower back and rump, with three broad black stripes; flanks, with a black spot, and region of tail crossed with a wide bar of the same."—*Ridgway.*

# AMERICAN AVOCET.

IRREGULARLY distributed throughout temperate North America, the Avocet is rather less abundant on the Atlantic Coast, but very numerous in the West, on the plains of Dakota, Montana, Wyoming, Colorado, Utah, and Southeastern Oregon, not above an altitude of 4,800 feet. It breeds in various localities on the eastern coast and throughout the interior in suitable places, as far north as the valley of the Saskatchewan, and south into Texas. It is probably the best swimmer among the waders, and it is also possessed of a compressed, thick, duck-like plumage impervious to wet or dampness. When not molested in their haunts, they are exceedingly tame and confiding, hardly paying any attention to the report of a gun, but, like other creatures, when much hunted become wild and wary. On taking wing the long legs are permitted to hang loosely down, until the bird is well under way, when they are stretched out stiffly behind to balance the long neck. Just before alighting they sail along for a short distance, and when on their feet the wings are elevated, as is the habit of various species of Plovers, etc., and then gradually tucked away beneath the feathers of the mantle and flanks. The food of the Avocet is insects and their larvæ, small crustacea, etc., and it is a beautiful sight to see a flock of these birds feeding. Wading along on the shallows the bills are moved regularly from side to side, through the water or mud, with the motion a man makes when mowing, each bird keeping

5. American Avocet.

to the side and a little behind the leader, and if the water is deep the head and neck are frequently immersed. They advance into the water up to their bellies, and if it should suddenly deepen they keep right on by swimming, not at all incommoded by the loss of their foothold. This species frequently utters a sharp *click*-like cry, and is often very noisy when on the wing, especially when disturbed from the nest. This is generally placed in grass in moist places, and is composed of twigs, grass, and sometimes seaweed. The eggs are four in number, varying in color from dark olive to buff, and spotted thickly with brown of various shades, such as chocolate, sepia, etc.

This species beside the name at the head of this article is called White Snipe and Blue Stocking.

*RECURVIROSTRA AMERICANA.*

*Habitat.*—Temperate North America, north to the Saskatchewan and Great Slave Lake, south to Guatemala, Cuba, and Jamaica. Rather rare in the eastern United States. Breeding from Great Slave Lake to Texas.

*Adult in Summer.*—Forehead and chin, rosy white; head, neck, and upper part of breast, light cinnamon; wings, terminal half of greater coverts, and inner secondaries, white; lesser coverts, inner scapulars, and near-lying feathers of the back, brownish black; primaries, brownish black, lighter on tip of inner web; entire rest of plumage, white, except tail, which is pearly gray; bill, black; legs and feet, pale blue; iris, bright red; webs, partly flesh color. Length, 15½–18¾ inches; wing, 8½–9; tarsus, 3¾; culmen, 3½.

*Winter Plumage.*—Same as above, except the cinnamon of head, neck, and breast is replaced with white, tinged with bluish gray.

*Young.*—Similar to the winter plumage, but the primaries are tipped with whitish, scapulars and back transversely mottled with buff, and hindneck tinged with rufous.

## BLACK-NECKED STILT.

ABUNDANT in certain portions of the Western and Gulf States, this long-legged wader is met with only occasionally in the eastern and northeastern section of our land, although specimens have been procured at Grand Menan, and near Calais, Maine. While its range is similar to that of the Avocet, with which it often associates, it is most abundant in the middle Western States and thence southward, but does not proceed so far to the northward as its ally. The long legs appear so slender as to be hardly capable of upholding the weight of the body, and the bird has the habit, just after alighting, of standing with half-bent legs and trembling wings, as though unable to keep itself erect, and this curious custom has doubtless given rise to the belief that the legs are too feeble to afford the necessary support. But the fact is, that the Stilt walks firmly and gracefully, the legs much bent at the heel at every step, as is only natural, and there is no more unsteadiness in its gait than is witnessed in other birds with but a quarter of its length of limb. Provided with long wings, that reach when folded beyond the tip of the tail, this species has a flight both swift and easily maintained, and it has the habit in its progress of exhibiting alternately the upper and under side of the body, like many other species of the *Limicolæ*, affording a pleasing contrast from the black of the back to the pure white of the under parts, brightened by the long lake-red legs extending beneath and beyond the tail. While

6. Black-necked Stilt.

amply provided with the means of seeking its food in shallow water, the Stilt is a poor swimmer, from the fact that its feet are not webbed, but when it finds itself in too deep water for wading, as occasionally happens, it progresses for a short distance without much difficulty, until able again to reach the bottom or the shore. This species goes in flocks, but frequently is seen only in pairs. The food consists of insects and their larvæ, small crustacea, worms, fry of fish, etc., which are deftly seized by the point of the sensitive bill, and in feeding often the head and part of the neck is plunged beneath the water. In the United States the Stilt breeds in the North from Southeastern Oregon and Great Salt Lake to the Gulf States. The nest is either a slight depression in the ground lined with grass, or else a sort of platform of straw and grass placed in the marshes, sometimes just raised above the level of the water. The eggs, three or four in number, are drab, brownish olive, occasionally rufous, for the ground color, blotched and spotted with brownish black, and measure 1.60–1.85 inches in length by 1.15–1.25 in breadth. It is stated that incubation is performed by both sexes, but I have not been able to corroborate this fact myself. The eye of the Stilt is large and beautiful, and the cry is a sharp *click*-like note, frequently uttered on the wing, especially when disturbed from its nest. The young run almost as soon as hatched. This species is also called Lawyer, White Snipe, Stilt, Tilt, Longshanks, and Tildillo.

*HIMANTOPUS MEXICANUS.*

*Habitat.*—Northern United States, southward to Peru on the west, West Indies and Brazil on the east. Rare in the Eastern States, except Florida Breeds from Southeastern Oregon and Great Salt Lake to the Gulf States.

*Adult Male.*—Forehead, spot above and below the eye, chin, throat, front and sides of neck, and entire under parts, together with rump and upper tail-coverts, pure white; remainder of head, hind-neck, back, and wings, glossy greenish black, brightest on the back and wings; tail, ashy white, bill, black; legs and feet, lake or rosy pink; iris, crimson.

*Adult Female.*—Differs from the male in having the back and scapulars brownish slate.

*Young in First Plumage.*—Have the feathers of the back, scapulars, and tertials bordered with buffy white, while the back of head and nape is mottled with the same color.

*Downy Young.*—Above, grayish white, mottled with dusky, with broad black blotches on back and rump; wings, rufous; head and hind-neck, grayish, mottled with black, and a vertical black stripe on head; under parts, white, washed with gray on fore-neck.

7. European Woodcock.

## EUROPEAN WOODCOCK.

THE European Woodcock is included in the list of North American birds simply from the fact that individuals are occasionally captured within our limits. It is much larger than our well-known species, frequently weighing fourteen ounces, and has a very differently colored plumage. In the Eastern Hemisphere it ranges throughout the northern parts from the Atlantic to the Pacific, and breeds as far north as the Arctic Circle, and south to the Azores, Madeira, in the Himalayas at 10,000 feet elevation, and in Japan. The winters are passed in the Mediterranean basin and in similar latitudes, as far east as China. Chiefly a nocturnal bird, it frequents the woods during the day and seeks the marshes and other suitable grounds in the evening to feed. In its habits it resembles our own bird and the food is similar. As a table delicacy it is highly esteemed in European countries, but in gastronomic qualities it is inferior to the American species. The sexes are alike in plumage, the female, as usual, being the larger. The nest is not much more than a depression in the ground, lined with a few leaves and grass. The eggs, four in number, vary from grayish white to brownish buff, with irregular reddish brown and ashy-gray spots. Where any unusually large woodcock is reported to have been killed within our limits, it is pretty certain to be this species.

*SCOLOPAX RUSTICOLA.*

*Habitat.*—Northern portions of Old World. Accidental in eastern North America.

*Adult.*—Anterior portion of crown and forehead, buff gray, with narrow, central, dark-brown lines; remainder of crown and nape, black, crossed by four narrow buff or pale rusty lines, a conspicuous dark-brown line from nape to eye; chin and throat, white, spotted with brown; neck all around, buff, crossed with fine dark-brown lines; upper parts reddish chestnut, vermiculated with buffy spots and brown lines, and blotched on the back and scapulars with black, the latter as well as the back mixed with light grayish; rump, reddish brown, barred narrowly with black, the upper tail coverts tipped with gray or buffy; tail, black, margined on outer web with chestnut and tipped with grayish buff on the upper surface, and with silver white on the lower; entire under parts, grayish buff, narrowly barred with brown; primaries, dark brown, transversely banded with cinnamon on outer webs; bill, flesh color, graduating into brown for the terminal third; legs and feet, flesh color; iris, hazel. Length, 13½ inches; wing, 8; culmen, 3¼; tarsus, 1½.

*Young* differ chiefly in having buff bases to the gray tips on the upper surface of the tail-feathers, and the chestnut on the outer webs becomes bars reaching to the shaft.

8. American Woodcock.

## AMERICAN WOODCOCK.

KNOWN familiarly to sportsmen and others throughout the country, this favorite game bird is gradually becoming scarcer within our limits. The high price it brings in the market, and the constant demand for it from the wealthy denizens of our cities, has caused it to be assiduously sought after by gunners in every locality where it was likely to be found, and from the time the young are scarcely able to fly until they depart on their migrations into places where, perchance, they may obtain a temporary refuge, their pursuit is never relaxed nor their slaughter discontinued. It is found generally throughout the eastern United States, but rare west of the Mississippi, and wherever found in summer, there it breeds. Although known to the majority by its name of Woodcock, it nevertheless has many aliases in the different parts of the country which it visits, and is called Big Mud, Big-headed, Blind and Wood and Whistling Snipe; Bog-sucker, Timber Doodle, Bog Bird, Night Partridge, Night Peck, Hookum Pake, Pewee, Labrador Twister, Whistler, and probably many others. Being a migratory species, the length of its stay in any particular locality depends greatly upon the weather, for though perhaps very abundant on one day, yet if during the night from sudden cold their feeding ground becomes frozen, by the next morning not a bird would be found, all having departed to a milder clime. It migrates always at night, when, indeed, it is most active, for it is a nocturnal bird, its sight being much better after the sun has departed than when the eye is exposed to the full light of day. In the Northern States it breeds in March, the eggs being often dropped when the ground is still covered with snow. The nest is a very slight affair, merely a depression in the ground, lined with leaves and grass, and formed in a secluded part of the woods, and the average number of eggs is four. They are of different shades of buff, thickly spotted with brown, and measure from 1.53–1.58 × 1.14–1.20 inches. The Woodcock resorts to the border of rivulets and margins of muddy ground, and procures its food by thrusting the bill up to the nostrils into the soft earth, and by means of the flexible extremely sensitive tip, seizes and draws out the worms which constitute its principal food, swallowing often as many in the course of the night as would equal its own weight. It also frequents the cornfields, and in the autumn is found upon the hillsides often at a considerable distance from any water, where it turns over dead leaves to seek for worms lying beneath. The eye of the Woodcock is large, bright, and beautiful, placed high upon the head, a position that protects it from injury when the bird thrusts the bill deep in the mire, at the same time enabling it, when thus employed, to see any approaching danger at a considerable distance.

The flight of this bird is rapid and erratic, and when flushed in its retreats it rises to the height of the surrounding bushes and almost immediately drops to the ground, upon which it runs for a short distance. No sport is more attractive than that of Woodcock shooting, although it has its drawbacks, such as the dense covert through which the gunner must force his way, his path beset by vines and interlacing branches that impede his progress and prevent him from seeing his faithful four-footed companion, who may at the very moment be standing rigid, as though carved in stone, but a few feet from him, with the hot scent of the close-lying bird welling up into his sensitive nostrils. Then, in the humid July days the heat is frequently oppressive in the low swamps, and too often the plaintive song of the mosquito is persistently crooned in the sportsman's ear, and he receives on every exposed part the sharp attentions of this never-dismayed insect. In the spring the Woodcock has some curious habits. With its bill inclined towards the ground, and with a forward movement of the body, it emits a sharp note resembling *ping-k* or *kwan-k*, and then rises in a spiral flight to a considerable height, with a shrill sound caused by the wings, and after flying in irregular and erratic circles at its greatest elevation, returns rapidly to the ground, uttering a sharp whistle, and alights near the spot from which it rose. When disturbed, on taking wing, a sharp whistling sound is produced, that, many claim, proceeds from the mouth of the bird, but this is difficult to prove, and it is more probable that it is caused by the wings, the falcate primaries cutting the air as they are moved by the rapid beat of the pinions.

Once I was watching a Woodcock in the early spring, as I remained concealed from its view in the midst of a swamp, and was surprised at its curious antics. After standing motionless for a few moments, as if in deep thought, it would suddenly draw its head close towards its back, with the bill pressing hard against the breast, drop the wings until the primaries touched the ground, and raise the tail with the feathers widely outstretched and elevated, so that the long under-coverts were plainly seen, and in this compressed attitude of a miniature turkey-cock, would strut about with quick elevation of the feet, as though the intensity of its feelings were altogether so overpowering as to preclude its walking upon such an ordinary every-day thing as common swamp mud. In a few moments the spasm seemed to pass, and coming back to the realization of its usual condition and surroundings, it would look about for a moment and again relapse into a contemplative mood. This action was repeated several times, but the bird uttered no sound.

The young are comical-looking little objects, covered with a yellowish down marked with black, and with a bill apparently much too long for them to successfully manage. They totter about as soon as hatched, immediately deserting the vicinity of the nest, and are not able to fly until nearly a month old. The female is very attentive to them, and often transports them about by clasping the little creatures with her legs and pressing them against her body. She also practices all the arts usual to birds, of counterfeiting lameness, or inability to fly, in order to distract the attention from her offspring of any intruder who has suddenly approached the helpless young, or nest with its complement of eggs, and entice him to follow her with the hope of a speedy capture, when, having drawn the object of her fears to a safe distance, she disappears on sounding wings into the recesses of the swamp. During the greater portion of the day the Woodcock remains concealed and quiet, its activity and searching for food commencing with the gloaming. In Georgia, North Carolina, and some other Southern States the Woodcock is resident throughout the year, but appears in the Northern States from its southern rambles in February, and begins at once to busy itself with the pleasures of courtship and its resulting duties.

## *PHILOHELA MINOR.*

*Habitat.*—Eastern North America, westward to the plains and north to the British Provinces. Accidental in the Bermudas. Breeding within its range.

*Adult.*—Head, rufous ash, with a dark-brown line from mouth to eye, and another across ear-coverts; occiput, black, crossed by three narrow buff lines; upper parts, variegated with ash, rufous, and black, the latter in large blotches on the back and wings, the ash frequently arranged in bars on the interscapular region; entire under parts, reddish buff, washed with gray on the breast and brightest on the flanks and under the wings; tail, black, tipped with ash above, and spotted on margin of outer webs with rufous and tipped beneath with white; under tail-coverts, with central lines of black; quills, dusky brown, first three narrow and somewhat sickle shape; bill, brown, yellowish at base of mandible and black at tip; feet and legs, pale reddish. Length, $10\frac{1}{2}$–$11\frac{3}{4}$ inches; wing, $4\frac{3}{4}$–$5\frac{3}{4}$; tail, $2\frac{1}{4}$; bill, $2\frac{1}{2}$–3; tarsus, 1.4.

*Downy Young.*—Head, grayish buff, with a line from bill to eye, top of head, extending irregularly down the hind-neck; spot behind the eye, connected with a larger one below it, brownish or chestnut black; back, blotched and vermiculated with black, grayish brown, and buff; a broad dark-brown stripe on lower back and rump; under parts, isabelline, washed with gray on the throat, reddish buff on flanks and belly, fading into pale buff on lower abdomen; wings, snuff brown feathers margined with buff.

## EUROPEAN SNIPE.

THIS is the common Snipe of the Old World, and in its habits does not differ from our Wilson's Snipe. It is found throughout Europe and Asia, as far south as North Africa, and eastward to the Philippine Islands. It breeds throughout Northern Europe and Siberia, but is rare north of latitude 70°. It has been found in considerable numbers at times in Greenland, and Reinhardt thought it bred there, but no eggs have been taken on that island. In Bermuda its appearance can only be considered as purely accidental. In North America it has never been met with, and its adoption into our avi-fauna can only be authorized by acceding the possibility of its passing from Greenland to our northern boundaries, but it is in no sense a North American bird. Its cry or note resembles *gick-jack*, *gick-jack*, quite different from the harsh *scaipe* of the American species. While the usual number of tail-feathers is fourteen, individuals are occasionally found with sixteen, and sometimes with only twelve. Pale or albinoid varieties, and also very dark or melanoid individuals, have been procured. These last have even received a distinct name, *Scolopax sabinii*, and the bird with sixteen feathers in the tail has been called *S. brehmii*, while a large rufous-colored individual was designated a *S. russata*. None of these, however, are worthy of separate recognition. The nest of this species is only a depression in the ground lined with grass, usually in a bunch of rushes, or amid the grass.

The eggs are four in number, varying from green to greenish buff, blotched with dark reddish brown, thick-

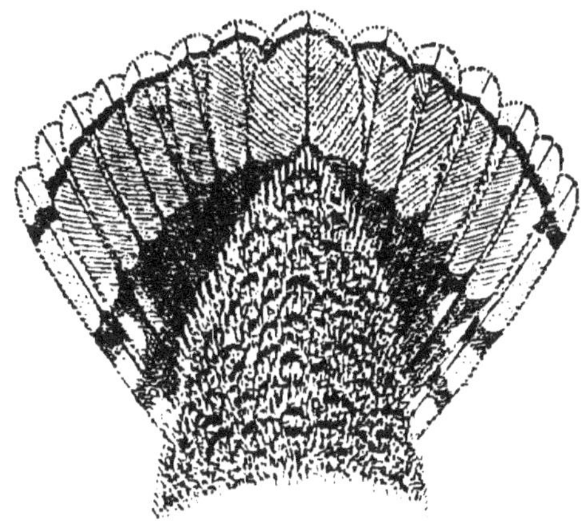

TAIL OF EUROPEAN SNIPE.

est at the largest end, and with underlying spots of purplish gray. They measure from 1.51–1.59 × 1.08–1.15. The shape is pyriform.

### *GALLINAGO GALLINAGO.*

*Habitat.*—Northern parts of the Old World. Occasional in Greenland. Accidental in the Bermudas. Breeding throughout Northern Europe and Siberia.

*Adult.*—Very similar to next species, Wilson's Snipe, in general appearance, but the tail usually consists of fourteen feathers; top of head, blackish, with central and lateral buff lines; chin and upper part of throat, whitish; neck all round, and upper part of breast, deep buff with dark-brown streaks; back, black varied with rufous and buff, the latter forming lines, and very conspicuous; wings, blackish brown, the coverts and secondaries barred and tipped with buff or buffy white; primaries, brownish black, and edged with white at tip, outer web of first pure white for nearly the entire length; rump and upper tail-coverts, rufous barred with black; tail, central feathers black, with a rufous bar on apical portion, and tipped with buff; remainder rufous barred with black; axillaries and flanks, white barred with black; rest of under parts, white; under tail-coverts, reddish buff barred with black; bill, reddish brown at base, dark brown at tip; legs, pale greenish; iris, dark brown. Length, 10 inches; wing, 5; bill, 2¾; tail, 2⅓. The sexes are alike; female the larger.

10. Wilson's Snipe.

# WILSON'S SNIPE.

HIGHLY esteemed as a game bird, the present species, known universally as the English Snipe, is distributed at different seasons of the year throughout the United States from the Atlantic to the Pacific. Besides the name given above, it has many local appellations, and is called in different sections Jack Snipe, Bog and Marsh Snipe, Alewife Bird, Shad Bird, Shad Spirit, Gutter Snipe, etc. Though occasionally nesting in different parts of the Northern States, its breeding range is from latitude 42° north to well within the Arctic Circle. In its migration southwards it penetrates into South America. In September it arrives in the United States from its summer quarters, and is found frequently in great numbers on the marshes, banks of rivers, or borders of ponds; in fact, in all suitable localities where its accustomed food of worms, insects, etc., can be procured. On taking wing this bird utters a harsh grating sound like *scaipe*, several times repeated, and frequently for a number of yards flies in a zigzag course, very discouraging to a young sportsman, who generally discharges his gun at the place the Snipe has just quitted. In a few moments this erratic course is changed and the flight continued in a direct line. If the day is still and warm, this snipe will lie very close, seeming averse to rise from the ground, and when compelled to take wing only goes a short distance before alighting; but if the weather is boisterous and the sky cloudy, it is often very wild, rising a long way ahead of the sportsman, and to a great height in the air, where it is joined by its fellows who have been startled from the same ground by the loud *scaipe*, *scaipe* of the first risen bird, and the flight is often continued for a long distance until the birds have vanished from sight. It is best to hunt them *down wind*, so that when the bird rises it will be against the wind and towards the sportsman, who will thus have a cross shot as the snipe endeavors to fly by him. It is easily killed, but when wounded hides so successfully (its plumage according so well with the color of the ground it frequents), that it is almost impossible to find it without the aid of a well-trained dog. Sometimes the report of a gun will cause numbers to rise from different parts of the marsh, and after executing various eccentric movements, often high in the air, the birds will pitch headlong with great velocity and alight near the spot from which they rose. Exhibitions such as this, after the sportsman has reached the marsh, do not, as a rule, foretell a successful hunt. It shows that the birds are uneasy and inclined to be wild, and in all probability after a few shots have been made, they will rise in a bunch to seek some other feeding place.

Over the muddy, often treacherous ground it frequents, the Snipe walks easily and lightly, carrying its bill slightly downward, and on reaching a suitable spot thrusts it several times in rapid succession up to the nostrils into the yielding mire in search of its accustomed food. Like that of the Woodcock, the tip of the bill is very flexible and sensitive, and the hidden worm is quickly felt, seized and drawn out. The Snipe is a voracious feeder, and one bird will cover quite a large extent of ground in a single night in its search for food, leaving as evidence of its diligence great numbers of small holes where the soil has been pierced by the busy bill. At times in the autumn this bird is found upon the uplands, at a distance from its usual marshy localities, and in such places it must procure its subsistence by searching for worms, etc., like the Woodcock, beneath the dead leaves. It begins to return from its southern haunts in February, and by May nearly all have passed onwards to their breeding grounds. When mated, the two birds perform curious evolutions upon the wing in the early morning, rising high in the air and sailing rapidly around each other, producing a strange rolling sound as they descend with great velocity to the ground. A similar maneuver called "drumming" is often witnessed, chiefly in the autumn when the birds are wild. One will rise to a great height and then descend towards the earth with the swiftness of an arrow, creating a sound (caused, as is generally believed, by the rush of air through the primaries) which can be heard for a considerable distance. This noise startles other birds in the vicinity until the air is filled with snipe, "drumming" in all directions, if they are numerous in the locality. This action on the part of the birds generally foretells a poor day's sport, as they will rarely lie at such times, but rise at long distances, and if followed soon leave the place. The arrival and departure of Wilson's Snipe is often very sudden and unexpected, as it will appear in great numbers in a locality one morning, where the previous day not a bird was to be seen, and disappear with equal celerity on some subsequent night. They travel always at night, preferring those usually when the moon is shining. Not infrequently this bird will alight on the topmost rail of a fence, or stump, or even on the branch of a tree, and it is perhaps more addicted to this habit when in the vicinity of its nest. This is merely a depression in the grass or bog. The eggs, about four in number,

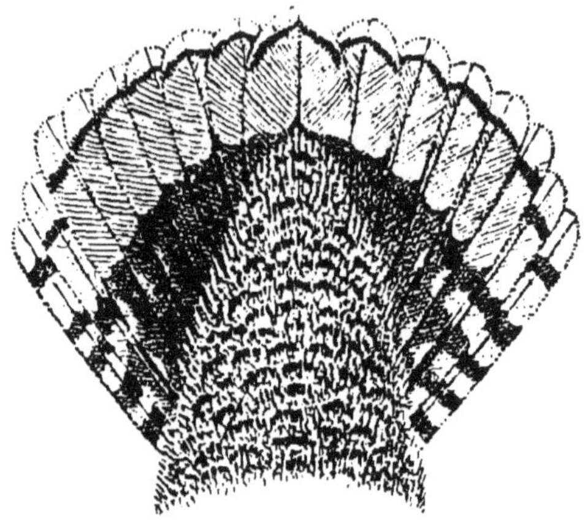

TAIL OF WILSON'S SNIPE.

placed with their small ends downwards, pyriform in shape, are grayish olive or olive brown in color, spotted and blotched with reddish brown, most numerous at the larger end. They measure 1½ inches in breadth by 1 1-10.

### *GALLINAGO DELICATA.*

*Habitat.*—Whole of northern and middle America. North to Hudson's Bay on the east, and Sitka, Alaska, on the west; south in winter to northern South America, and to the West Indies. Breeding from latitude 42° northward.

*Adult.*—Top of head, black, with median and lateral stripe over each eye, buff; neck, buff, with numerous fine black spots or lines; back, black, feathers barred with rufous and margined with pale buff, arranged conspicuously in long lines; rump and upper tail-coverts, rufous, barred with black; wings, brownish black, feathers barred with rufous and margined with white; primaries, blackish brown, with web of first pure white for nearly its entire length, edged with white at tip; tail, usually of sixteen feathers, all but three outer on either side, black with sub-apical rufous bar, and tipped with whitish buff, the three outer pale buff or whitish, with narrow black bars; chin and upper part of throat, buffy white; lower part of throat and breast, buff, spotted with brown or brownish black; flanks, white, barred with black; abdomen, pure white; under tail-coverts, buff, barred with brownish black; bill, greenish, yellowish at base of mandible; legs and feet, greenish. Length, 10½–11½ inches; wing, 5–5½; tail, 2¼; bill, 2½–3.

## DOWITCHER.

ON the Atlantic seaboard, where it is called in various localities, the Red-breasted Snipe, Dowitch, Gray and Brown Snipe, Brown-back, Grayback, Quail Snipe; and Robin Snipe in Connecticut, this species is one of the most common and well known of the "Bay-birds." By the majority of writers it has been confounded with the next species, the *Long-billed Dowitcher*, but from which it is specifically distinct, and in the various published biographies of the two birds, the ranges are considerably mixed. The present species is an inhabitant of eastern North America, although undoubtedly examples are found at times in different parts of the continent, but the two species are essentially the eastern and western representatives of the genus, and their migrations are performed on different lines of travel, the present species confining itself mainly to the eastern littoral, though it has been witnessed in Illinois when migrating through the Mississippi valley, as stated by Cooke (U. S. Dep. Agriculture, Bull. 2, 1888, p. 92). This is probably its most western migratory limit in the Middle States. Messrs. Dall and Bannister state that they found this species not common at the mouth of the Yukon River, and at Pastolik. It is an extremely gentle, sociable bird, goes in small flocks, the individuals of which keep closely together and perform their various graceful evolutions when on the wing, as if moved by one common impulse.

12. Dowitcher.

It arrives from the south on its way to the far northern breeding grounds in April, but does not tarry, and returns in August, when it proceeds leisurely, stopping at every suitable place, to seek its food and rest from the fatigues of the journey. At such times there is no need for haste. The duties of incubation and care for the young have all been fulfilled, and the little wanderers take life as easily as possible and enjoy it to its fullest extent, as they proceed towards the land of the magnolia and myrtle, to escape from the harsh, inclement season of northern climes. When a flock arrives at some sand-bar, muddy flat, or meadow on our coast, the individuals composing it, after circling around as if selecting the most attractive spot, alight in a body, stand motionless for a few moments, and then disperse over the ground to seek their food, all the time keeping up a pleasant low chatting, as though carrying on an extended conversation with each other. Should any dangerous intruder draw near, they are among the last of the birds in the vicinity to take alarm, following their peaceful occupations frequently until it is too late to escape, and pay as a penalty for their confiding disposition the forfeiture of life. To witness a flock of these Snipe come to the decoys is a beautiful sight. The gunner, hidden in his blind of grass and reeds, with his wooden counterfeits of various species of waders strung out before him, notices a dusky mass at some distance drawing towards his retreat ; and by the manner it moves along, he recognizes that a flock of "Dowitchers" is approaching, and commences to whistle an imitation of their note. The birds soon hear the well-known sound, and begin to look for the spot where their supposed relatives are feeding. Soon they detect the decoys, and in compact mass wheel towards them, in gentle tones replying to the false call. Swinging above the wooden images, they hover for a moment preparatory to alighting, when the ambushed sportsman rises, and by quick discharges from his gun, hurls his deadly missiles into their crowded ranks, strewing the ground with the dead and wounded. The survivors, startled by the reports and the vacancies in their ranks, rise and rapidly flee from the dangerous place, but have proceeded but a short distance, when again their familiar cry is borne to them on the air, and they wheel again, and unsuspicious and confiding as ever, return once more to meet their supposed friends, and once more are received with deadly missiles. It is not unfrequently that in this way an entire flock has been destroyed, so solicitous seem the unwounded birds to remain with their fellows. At times, instead of flying over the decoys, they will alight a short distance away, and huddled together in a dense, compact mass, as though each gained confidence from the close proximity of his neighbor, they stand motionless, and are mowed down by the remorseless gunner. If, however, they should have alighted out of gunshot, in a little while they seem to regain their usual confidence, and the birds will scatter over the ground, intent only upon seeking their food, in the pursuit of which some may again approach the decoys, uttering their low, gentle note as they draw near the fatal ambush.

Dowitchers associate on most friendly terms with other species of waders, such as Yellow-legs, Willets, and the various kinds of Sandpipers. Their flight is rapid and is often greatly protracted. When the birds are traveling at a high elevation, either for the purpose of seeking new feeding grounds or distant localities, they are difficult to decoy, and pay but slight attention to the imitation of their cry, be it whistled ever so accurately. The Red-breasted Snipe breeds in the far North, and its eggs have been taken near Fort Anderson, in the Arctic regions. The nest, placed on the marshy shores of lakes, in a depression in the moss, was lined with a few leaves and grasses. The number of eggs was four, of a drab or fawn color, with shadings of rufous or olivaceous, covered with chocolate or sepia markings, most numerous at the larger end, and measure 1½ to 1¾ by 1 1-10 to 1 1-5 inches. As may be supposed from its partly web-foot, this species swims fairly well when necessary, keeping time to the stroke of its legs by a backward and forward movement of the head and neck, and when wounded is very skillful in hiding in the grass.

In the British Islands straggling individuals of this species have been taken at various times, and it has also been killed in France, near Havre, and in Picardy.

## *MACRORHAMPHUS GRISEUS.*

*Habitat.*—Eastern North America to the Arctic regions; casually in Alaska (Nushagat River). In winter south to the West Indies and South America. Occasionally in Bermuda. Straggler to the Old World. Breeding in Arctic regions from eastern seaboard to Rocky Mountains.

*Adult in Summer.*—Head and upper parts, mixed with buff, rufous, and white; lower part of back, white; rump and upper tail-coverts, white, barred with black; wing-coverts, grayish brown, margined with white; secondaries, black, barred with rufous; primaries, brownish black; shaft of first primary, white; throat, front and sides of neck, breast, and upper edge of flanks, cinnamon, spotted with dark brown, barred with same on flanks; abdomen, white, tinged with buff; under tail-coverts, whitish buff, barred with blackish brown; tail, white, barred with black, the central pair sometimes tinged with buff; bill, blackish brown; legs and feet, greenish brown. Length, 10–12½ inches; wing, 5⅝; culmen, 2¼; tarsus, 1½; middle toe, 1 inch.

*Adult in Winter.*—General plumage, ash gray, mixed with white on breast and flanks; stripe over eye, lower part of back, abdomen, and belly, white; wing-coverts, margined with white; wings, darker than rest of plumage, but without any buff or cinnamon markings; tail and its coverts, as in summer, but without any buff tinges.

13. Long-billed Dowitcher.

## LONG-BILLED DOWITCHER.

THE Long-billed Dowitcher, also called Long-billed Snipe, Western Red-breasted Snipe, Greater Long-beak, Greater Grayback, Western Dowitcher, Red-bellied Snipe, and Jack Snipe at Los Angeles, and White-tail Dowitcher on Long Island, is, as its various names imply, a somewhat larger bird than its relative, the Dowitcher of the Atlantic Coast, and with a longer bill. It has been usually considered by authors as a variety of, or identical with, its ally, and all the records of *Macrorhamphus* from the Mississippi Valley to the Pacific can with tolerable certainty be ascribed to this species. It has been found common at the mouth of the Yukon by Dall, but although rare in the breeding season, was observed in large flocks in August at Cape Smythe, Alaska, by Murdoch. On the American coast of Behring Sea it is very common; is also found on Kotzebue Sound, and even farther north, and breeds throughout this range. Its appearance on the Atlantic Coast is at very irregular intervals, and although quite a considerable number of specimens have been obtained at various times, it may be regarded as an accidental straggler, individuals having probably joined a flock of the eastern Dowitchers and accompanied its members along the Atlantic Coast, instead of proceeding, as usual, to the westward. Nelson, who had opportunities of observing this species in the breeding season, on the shore of Norton Sound and the banks of the Yukon, in Alaska, says it arrives at its summer home from the 10th to the 15th of May, when the snow disappears. At the end of the month they are numerous, and the most conspicuous frequenters of the marshes. Their method of love-making is very energetic. Two or three males give chase to one female and pursue her over the marshes, twisting and turning in their flight with wonderful dexterity and swiftness. Occasionally a male checks his flight and utters a shrill *péet ŭ wéet; wée too wée too;* then on again at full speed. When mated, or but one male is paying his court, the two rise in the air for a short distance, and hovering for a few moments, the male utters an energetic and frequently a musical song, something like *péet-peet; peé-ter-wée-too; wée-too;* twice repeated. The nest is merely a depression in the moss, without any lining, and the eggs are usually four in number, with the ground color, greenish olive, light gray, or clay color, covered with large umber-brown spots, most numerous at the larger end.

The young are full grown and on the wing by the last of July, and by the first of September the full winter dress has been assumed, and they frequent muddy flats and shores of tidal creeks, as is their custom when performing their migrations. They are as unsuspicious and gentle as their relatives, and have the same reluctance to leave their wounded companions who have fallen at the discharge of the gun, and may be shot at frequently before the survivors compel themselves to leave the dangerous neighborhood. On the Pacific Coast of the United States, south of British Columbia, it is found in the autumn very abundant about the lagoons and banks of rivers, feeding in water that their legs would just permit them to wade in, and probe the bottom with their long bills. Specimens have been obtained of this species in Cuba.

### *MACRORHAMPHUS SCOLOPACEUS.*

*Habitat.*—Mississippi Valley and western North America, from Alaska to Mexico. Accidental on the Atlantic Coast and in Cuba. Breeding on both coasts of Behring Sea and in Alaska.

*Adult in Summer.*—Top of head and back of neck, cinnamon, streaked with black; buffy-white line from bill to above the eye; loral space covered by a broad dark-brown bar; back and wings, black, feathers margined with reddish and white; coverts, dark brown, margined with white; primaries, blackish brown, shaft of first one white; lower back, pure white; rump and upper tail-coverts, white, barred with black; throat, pale buff; front and sides of neck, cinnamon, spotted with brown; entire under parts, uniform cinnamon, palest at vent; edge of flanks and under tail-coverts, barred with black; central tail-feathers, black, barred with white and pale buff, remainder blackish brown, barred with white; bill, longer than that of *M. griseus*, and together with legs and feet, is blackish green. Length, 10¾ inches; wing, 6; culmen, 2¾; tarsus, 1⅝; middle toe, 1¼.

*Adult in Winter.*—Head, back, and wings, dark gray, mixed with dark brown, and margined with whitish brown on wing-coverts; lower back, rump, tail-coverts, and tail, as in summer, but without any buff; stripe from base of culmen extending over and behind the eye, abdomen, and vent, pure white; throat, white, faintly streaked with dusky; neck in front and sides, together with the breast, brownish gray; flanks and under tail-coverts, pure white, barred with black.

## STILT SANDPIPER.

FORMERLY considered as among the rarest of our waders, the Stilt Sandpiper is still one of the little-known members of the family, and is generally met with singly or in small parties of five or six individuals. It associates with the little Yellow-legs and the Dowitchers, and although very gentle, is rather more wary than either of those species. I have met with it on the Jersey coast, near Barnegat, at various times, mostly singly, or in small parties of three or four, but once in May a flock of two or three dozen came to my decoys. They flew with rather open ranks, and on alighting, the individuals kept a slight distance apart. After remaining motionless for a few moments, as is the custom of most waders, they began to seek their food, inserting the bills into the muddy soil. The legs appeared curiously long for the size of the bird, but they walked gracefully and with some dignity. They uttered occasionally a *tweet*-like note. After watching them for a short time I fired, when the survivors rose and made a rapid circuit of the flat, and again approached the decoys, hovering slightly over them and more bunched together, when I killed several ; the rest immediately started off and were soon out of sight. It was the only occasion that I saw so many together. It is stated that this bird, called on Long Island the Long-legged Sandpiper, will wade in water nearly covering the tarsus, and with the bill immersed almost to the base, sweep it from side to side, seeking for food somewhat

14. Stilt Sandpiper.

like the Avocet. I have never seen them do this, the method of feeding of those I have watched being that already described, and although there were numerous small ponds scattered about, they did not go near them, but kept upon the muddy flat from which the tide had but lately receded. Seebohm ("Charadriidæ," p. 401) says that this species breeds in the Arctic regions of America, from the Rocky Mountains to Baffin's Bay, but gives no authorities for the statement. McFarlane found it breeding at Rendezvous Lake, and it was tolerably abundant at Franklin Bay. He also obtained the eggs, which are now in the Smithsonian Institution. They are light grayish white, marked with dark brown, most numerous at the larger end, and measure 1.47–1.50 inches in length by 1 inch in breadth. The nest was merely a depression in the ground lined with leaves and grasses. I have not seen any record of its nest having been discovered in the western part of the Arctic regions.

## *MICROPALAMA HIMANTOPUS.*

*Habitat.*—Eastern North America, from the Arctic regions to the Bermudas, West Indies, Central America, Brazil, and Peru in winter. Not found on the Pacific Coast of the United States. Breeds from the Rocky Mountains to Baffin's Bay.—*Seebohm.*

*Adult in Summer.*—Front and top of head, black, streaked with buffy white and edged all around with rufous; line from bill to eye and ear-coverts, rufous; neck, white, tinged with buff and streaked with dusky; back, black, feathers edged with buff or white; wing-coverts, brownish gray; primaries, brown, darkest on the outer web, shaft of first one all white, of the remainder white for the terminal third; rump, dark gray; upper tail-coverts, white, barred with blackish; middle tail-feathers, pale gray; remainder, central portions white, margined with pale brown; throat, whitish, streaked with dusky; rest of under parts, dull white, slightly tinged in places with buff, and barred with dark brown; bill, black; legs and feet, greenish. Length, 7½–9¼ inches; wing, 5–5½; culmen, 1½–1¾; tarsus, 1½–1¾; middle toe, ⅞, and claw, 1 inch.

*Adult in Winter.*—Top of head, back, and sides of neck, loral streak and auriculars, gray; back, ash gray; wings, upper tail-coverts, and tail, as in summer; superciliary stripe and under parts, white, streaked with gray on neck, breast, and lower tail-coverts.

*Young.*—Top of head, brownish, streaked with buff; hind-neck, ash gray; back and scapulars, black, feathers margined with white, except those of the mantle, which are bordered with reddish; wing-coverts, margined with pale buff; upper tail-coverts, nearly pure white; lores, brown; throat and sides of head, white, faintly streaked with dusky; breast, grayish white, streaked with buff and dusky; rest of lower parts, pure white.

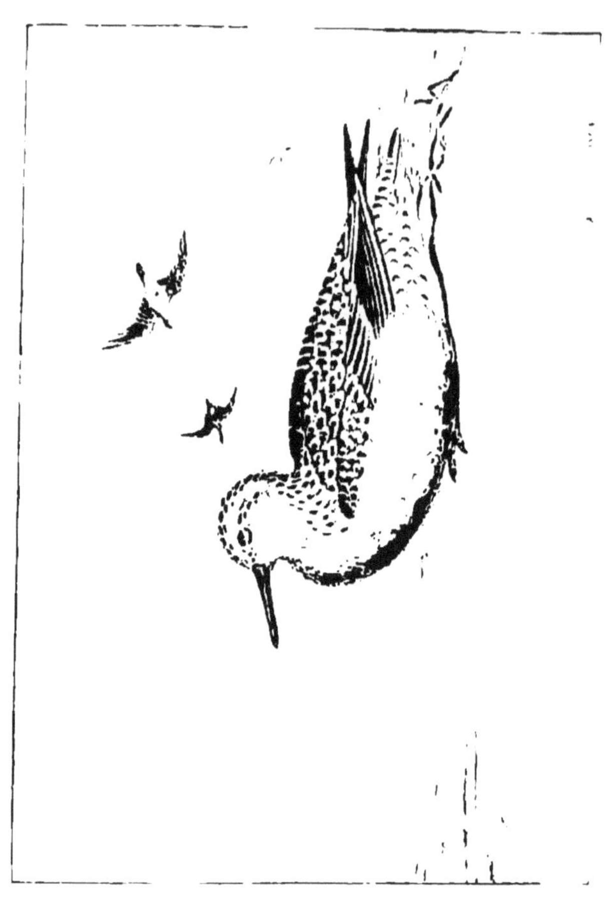

## THE KNOT.

LARGEST of all the Sandpipers, the Knot is found throughout the world, going in winter from its far northern breeding place to New Zealand, Africa, and Brazil. In the United States it is known by many names besides the one at the head of this article, a few of which are: Red Sandpiper, Ash-colored Sandpiper, Freckled and Grisled Sandpiper, Gray Back, May Bird, Robin Snipe, White Robin Snipe, White-bellied Snipe, Silver-back, Red-breast Plover, Buff-breast Blue Plover, Beach Robin, Robin-breast, and Horse-foot Snipe. In the spring, on its way north, and again in the autumn, the Robin Snipe visits the Atlantic Coast of America in great numbers, associating in flocks and remaining only a comparatively brief period in any one locality. On the seashore it follows the movements of the waves upon the beach, exhibiting great quickness in eluding the incoming surf, and also in following the retiring waters, rapidly picking up the aquatic insects and small bivalves left upon the sand. Although apparently wholly intent upon securing its food, it nevertheless keeps one eye upon the sea, and avoids with remarkable agility the tumbling surf that suddenly rises and breaks in front of it. The flight is performed with firmnes and speed, the birds indulging at times in many evolutions over both sea and land, executing these with much swiftness and remarkable unanimity. It also frequents the salt marshes and the many ponds and pools common to such tracts, and feeds on the minute shellfish found in the shallow water. At times it is very gentle and easily decoyed, approaching its wooden counterfeits without hesitation and paying with its life such misplaced confidence. At other times, more especially, perhaps, in autumn, during its migration southwards, it is more wary, and frequently declines to pay any attention to the sportsman's lures. It walks easily and runs with swiftness, and often probes the wet sand and muddy flats in search of its especial food.

At Point Barrow, Alaska, it was rather rare, but Murdoch was of the opinion it bred there, as a female was killed with full-sized eggs in her ovaries; but he never found the nest. Captain Fielden observed this species in considerable numbers in Grinnell Land, and soon after its arrival in June the courtship commenced. He saw two males chase a female. At this time they soar, like the common Snipe, and when descending, beat their wings rapidly behind the back, producing a whirring noise. At the mouth of the Yukon, Dall found it rare, and he obtained a young bird at St. Michael's. It appears to be more numerous about Hudson's Bay and on Melville Peninsula. Its food consists of small mollusks, worms, crustaceans, aquatic insects, and larvæ. The eggs of this species have for a long time been one of the greatest desiderata to oölogists. Parry found it breeding on the North Georgian Islands, and Richardson says Captain Lyon found its nest and eggs on Melville Peninsula, and the same writer states that the Knot breeds on Hudson's Bay down to the fifty-fifth parallel, but no specimen of the egg was in any collection until Lieutenant Greely brought one to Washington, obtained in the vicinity of Fort Conger, latitude 81° 44′ N. It measured 1.10 by 1 inch, and as described by Merriam was "light pea green in color, closely spotted with brown in small specks about the size of a pin's head." The bird was also captured at the same time. Captain Lyon states that the complement of eggs is four, and that the nest was only a tuft of withered grass, on which the eggs were deposited, the bird not taking the trouble to make any regular nest.

Some believe that when the breeding plumage has been once assumed it is never changed, and cite as proof the appearance in autumn of individuals with the red breast, and also imagine that it takes three or four years before the full dress is completed. I have seen more individuals which I consider fully adult, in the gray plumage, than I have in the breeding dress, and regard the evidence obtainable is against the view of an unchangeable dress.

### *TRINGA CANUTUS.*

*Habitat.*—Cosmopolitan. Throughout the seacoasts generally of both hemispheres, but not found on the Pacific Coast of America south of the Alaskan Peninsula. Abundant on the Atlantic Coast, rare in the Mississippi Valley. In winter to the West Indies, Trinidad, and Brazil. Breeding in Arctic regions.

*Adult in Summer.*—Head and upper parts, pale gray, variegated with black and reddish; rump and upper tail-coverts, white, barred with black; superciliary stripe, throat, fore-neck, breast, and sides of abdomen, light cinnamon; middle of abdomen, pure white; under tail-coverts and flanks, white, barred or spotted with black; bill, legs, and feet, black. Length, 10 inches; wing, 6½; tail, 2½; bill, 1½; tarsus, 1¼.

*Adult in Winter.*—Top of head and nape, dark brown, streaked with white; back and scapulars, ashy gray, with a subterminal bar to the feathers, and white tips; wings, rump, upper tail-coverts, and tail, as in summer; under parts, white; sides of face, neck, and breast, spotted and barred with ashy gray.

## PURPLE SANDPIPER.

THE Purple Sandpiper, called also, in Maine, "Winter Snipe," is a boreal species breeding in the far North and only coming in winter to the Great Lakes and Middle States of the Atlantic seaboard. It is not found on the shores of the Pacific. On Long Island it is not common, but abundant in various parts of the New England coast. It breeds on the Faroe Islands, Iceland, Greenland, and, according to Richardson, on Melville Peninsula and the shores of Hudson's Bay. In Europe it straggles south to Gibraltar, and in a single instance one is known to have gone as far as South Africa. It prefers a bold coast, and seeks its food amid the spray of the waves as they dash over the shelving rocks. Sure-footed, it clings easily to the slippery surface, and appears to delight in the angry waters and the commotion they cause around it. It is very gentle, allowing a near approach as it busies itself along the shore, merely running ahead for a short distance if the observer comes too close. It is said to be an excellent swimmer and will even alight on the water, but this I have never seen it do. At high tide it remains quietly on the rocks, pluming itself or resting, but becomes active when the water begins to recede. Occasionally it frequents marshy flats, in company with other waders, and subsists upon the minute organism peculiar to such localities. Its food consists of insects, mollusks, and small crustaceans, and the seeds of different plants common to the coast. Its flight is performed by rapid

16. Purple Sandpiper.

beats, with the wings curved downwards, and is not continued for any very great distance when near or on their feeding grounds. The breeding season begins in May, and the nest is placed on some rising ground in but a slight depression lined with dried moss or grass, and the eggs, usually four in number, are pale brownish buff, mottled and blotched with reddish or blackish brown, with underlying marks of violet gray or brownish gray, most numerous at the larger end. They measure about 1½ inches in length by 1 inch in breadth. The female employs all the usual artifices to lure an observer away from its nest or helpless young, feigning lameness, inability to fly, etc., tumbling over with almost every movement, and apparently has no thought for her own danger if mayhap she can secure the safety of her treasures. But one brood is raised in a season, and both birds appear to perform the duties of incubation. The note is loud and shrill, and the species has the habit of running along the shore with both wings elevated over the back. In summer it goes to the extreme North, as far as it is possible to procure any subsistence amid the ice and snow of those bleak and cheerless regions. In the interior of the United States the Purple Sandpiper visits the Great Lakes, is not uncommon on the shores of Lake Michigan, and has been noted as occurring in Missouri.

### *TRINGA MARITIMA.*

*Habitat.*—Northern portions of Old and New Worlds; most common in the northeastern part of North America, and southward in winter in the Mississippi Valley, and along the Atlantic Coast to Middle States. Casual in Florida. Replaced in the Alaskan Peninsula by an allied form. Breeds across Arctic America.

*Adult in Summer.*—Top of head, neck, back, and scapulars, blackish brown, feathers margined with chestnut, pale buff, or white; wings, grayish brown, coverts margined with white, forming a bar across the wing; some of the inner secondaries are mostly white; rump, upper tail-coverts, and central tail-feathers are brownish black, lateral tail-feathers light or grayish brown; obscure white streak from bill to above the eye; throat and fore-neck, white, streaked with brown; breast, grayish brown, tinged with rufous, the feathers tipped with white; rest of under parts, white, streaked on flanks and under tail-coverts with pale brown; bill, dark brown, paler at base; legs and feet, dull yellow; iris hazel. Length, 8 inches; wing, 4¾; tail, 2¼; tarsus, ⅞; culmen, 1⅛.

*Adult in Winter.*—Upper parts, black, glossed with purple, the feathers margined with gray; head, plumbeous; wings, as in summer, except they are somewhat darker and there is no chestnut on scapulars; throat, white; breast, lead color, tinged with purple; under parts, white, streaked on flanks and under tail-coverts with dusky brown.

*Downy Young.*—"Above, hair brown, lighter and more grayish on the nape, the brown irregularly marbled with black; the wings, back, and rump, thickly bespangled with whitish downy flecks on the tips of the down-tufts; head, pale rufous, variously marked with black; the crown, deep hair brown, variegated with black; beneath, entirely grayish white."—*Baird, B., and R.*

17. Coues' Aleutian Sandpiper.

# COUES ALEUTIAN SANDPIPER.

A WESTERN representative of the Purple Sandpiper, this bird bears so close a resemblance to its eastern ally, that it offers really no recognizable characters to definitely separate it. But not having at my command a sufficient series to enable me to reach an absolute decision, I have left this form, together with the Prybiloff Sandpiper, under a separate designation instead of reducing them both to synonyms of the Purple Sandpiper. I am under a strong impression, however, that all three represent but a single species. We learn from those who have met with this bird in its native haunts, that it is a common resident along the entire Aleutian chain of islands, and strays northward in the autumn, throughout the coast of Behring Sea. Dall found it at Nulato and Pastolik, and it is not uncommon at Sitka. It is also found on Kurile Island and the Siberian coast. It appears to breed from the most western of the Aleutian Islands to the Shumagin group south of Alaska. Nelson found it on Sanak Island of the Aleutian chain, where a pair were feeding on the rocks, uttering a note something like *clū-clū-clū*, and when on the wing it had a low clear, rather musical cry, like *tweo-tweo-tweo*. In August it was abundant at St. Michael's, in Norton Sound, going in flocks of from five to forty individuals, and frequenting rocky islets and rugged portions of the shore. When the snow began to fall in October, they retired to the inner bays and beaches, and were very tame, and permitted a near approach even after having been fired at. It is called by the natives "Beach Snipe" or "Shore Bird." In the Commander Islands it is found during both summer and winter, and in the latter part of March flocks of several hundred individuals are found upon the shore. Later these swarm over the island and settle in pairs, and commence the season of love-making, expressing their ardent feelings with a song. This is uttered on the wing, and is described as a loud, agreeable, and melodious twitter, the performer remaining suspended in the air meanwhile on quivering wings, and on its conclusion descending obliquely to the ground, where, perched upon some tussock, and apparently in a high state of excitement, it produces a "bleating" sound like that of the European Snipe. The eggs are laid about the middle of May, and are pale olive buff or brownish buff, spotted and blotched with umber brown. How far south on our west coast this bird goes in winter is not known.

### *TRINGA MARITIMA COUESI.*

*Habitat.*—Aleutian Islands, coast of Alaska, north to the Kowak River; west to Commander Islands, Kamtschatka.

*Adult in Summer.*—Head, neck, back, scapulars, and interscapulars, sooty black, streaked with reddish on the first two, and the feathers margined with reddish, buff, or white on the remainder; wings, dark grayish brown, lesser coverts margined with gray, greater coverts with white, forming a bar across the wing; rump, upper tail-coverts, and central tail-feathers, brownish black, rest of tail-feathers pale brown; white stripe over the eye, going to the nape; throat, neck, and under parts, white, streaked and clouded with brownish black, or sooty, sometimes mottled with buff on lower part of fore-neck and sides of breast, which latter is frequently all bluish gray, sometimes blotched and streaked with blackish on a white ground; flanks and under tail-coverts centrally streaked or terminally spotted with black; bill, blackish gray, yellowish at base; feet and legs, olive yellow. Length, 7½–9 inches; wing, 4½–5; culmen 1–1¼; tarsus, ⅞–1 inch.

*Adult in Winter.*—In this plumage Coues Aleutian Sandpiper resembles the Purple so closely that it is almost impossible to give recognizable characters to distinguish them apart. With specimens of both before me, beside a slightly shorter and more slender bill, a most unreliable character in waders, the present form is of a more grayish hue upon the back, with much less of the purple luster seen in examples of the Purple Sandpiper, and the feathers have rather lighter margins. In other respects there is little or no difference, and in a large series these variations would probably be of no value.

*Downy Young.*—"Above, bright rusty fulvous, irregularly mottled with black, the back, wings, and rump ornamented by yellowish white downy flecks or papillæ; head above, deep fulvous brown, with a longitudinal stripe of velvety black from the forehead to the occiput, where it is confluent with a cross band of the same; the lores with two nearly parallel longitudinal streaks of black; there are also other rather indefinite black markings, chiefly on the superciliary and occipital regions; lower parts, white, becoming distinctly fulvous laterally."—*Baird, B., and Ridg.*

## PRYBILOFF SANDPIPER.

AT first sight this Sandpiper seems very distinct from the two preceding forms, but on an examination of specimens the differences do not appear to be of any material consequence. It is somewhat lighter in color, and that statement comprises about all it can claim to a distinct specific, or even subspecific, rank. It is not even an island form, as was originally supposed, but goes in winter to the mainland of Alaska, having been taken in December or January at Portage Bay, on the Chilcat Peninsula, where it stays all winter in small numbers, appearing in flocks in spring, and it probably is distributed along the coast southward during the inclement season. It has been found on the Fur Seal Islands, and on St. Matthew's and St. Lawrence Islands. It reaches the Seal Islands in May, and breeds on the uplands and mossy hummocks. The nest is merely a depression in the moss, and the eggs, usually four in number, are olive yellow, with large and numerous markings of dark umber brown scattered over the surface, nearly confluent at the larger end, with underlying purplish-gray markings. The young are able to fly during the first week in August, and both old and young gather together in flocks and remain about the shores until September, when they all depart to their winter home. Like its relative, during the breeding season it utters a rather musical song composed of liquid notes, while suspending itself with rapidly beating tremulous wings a short distance above the ground.

18. Prybiloff Sandpiper.

The song finished, it descends and takes up its position on some projecting rock. In its flight and general habits, it does not seem to differ in any way from its relatives.

## *TRINGA MARITIMA PTILOCNEMIS.*

*Habitat.*—Breeding in the Prybiloff Islands, Alaska, Portage Bay, Chilcat Peninsula, and migrating to coast of adjacent mainland south of Norton Sound.

*Adult in Summer.*—Top of head, brownish black, streaked with buff; nape, pale fulvous, narrowly streaked with dark brown; loral stripe and ear-coverts, grayish fulvous, streaked with brown; rest of head and stripe over the eye, white; back and scapulars, black, the feathers margined with fulvous, reddish, or white; wings, grayish brown, feathers margined with white, forming a white bar below the greater coverts; some of the inner secondaries, pure white; primaries, pale brown, darkest on outer web; shafts, white; lower back, rump, upper tail-coverts, black, with a purplish shade on the coverts; middle tail-feathers, pale brown, the lateral ones graduating from a pale brown to grayish white as the outermost is reached, margined with a purer white; throat and sides of neck, white, indistinctly streaked with buff on the latter; rest of under parts, white, mottled and blotched with black on the breast, the blotches sometimes coalescing and covering all the breast; in some specimens there are indistinct central brown shafts on the flanks; bill (skin), brownish black; feet and tarsi, black. Length, $9\frac{1}{2}$-10 inches; wing, $5\frac{1}{6}$; culmen, $1\frac{1}{3}$; tarsus, 1 inch.

*Adult in Winter.*—"Wings, rump, tail-coverts, tail, and posterior lower parts, as in the summer plumage; remaining upper parts, continuous light ashy plumbeous; the feathers of the back and the scapulars, darker centrally, and with a very faint purplish gloss in certain lights; head, light grayish, darker and almost unbroken on the pileum, lighter and streaked with white elsewhere; the throat, white, and but sparsely streaked; jugulum and breast, white, irregularly marked with pale ash gray."—*Baird, B., and Ridg.*

*Downy Young.*—Paler than those of *T. M. Couesi*, but otherwise very similar; black loral streak not reaching the eye.

## SHARP-TAILED SANDPIPER.

NELSON appears to have been the first to introduce this species to the North American fauna, having obtained a female at St. Michael's in September, 1877. Later, others were seen, and in succeeding autumns it was one of the most common species, frequenting pools and tidal creeks, generally associated with the Pectoral Sandpiper, and combining together in flocks of from ten to fifty. They had a twisting flight, and pursued an erratic circuitous course, usually settling near the spot from which they started. On the Siberian coast, near North Cape, this species was very common, and resorted to the damp grassy flats near the shore, and sought its food in the tracks, made by the reindeer, which intersected the ground. It was here very tame and allowed one to approach closely. On Behring Island Stejneger obtained young specimens as they were migrating in autumn, but it does not appear to travel southward along our coasts. They were very shy, mostly single, no large flocks having been seen. In appearance this species resembles the Pectoral Sandpiper, but has never the streaked breast so conspicuous on the latter, and the top of the head is more reddish.

*TRINGA ACUMINATA.*

*Habitat.*—Alaskan coast and Eastern Asia, islands of the Pacific and Behring Sea, southward to Australia.

*Adult in Summer.*—Top of head, ear-coverts, and nape, streaked with black and rust red; white stripe from bill over the eye; back, scapulars,

19. Sharp-tailed Sandpiper.

and interscapulars, black, feathers margined rather broadly with reddish buff or rusty, grayish brown or white, back lighter than scapulars; lesser wing-coverts, black, margined with dark buff; greater coverts, dark brown, margined with white, forming a bar across the wing; primaries, dark brown, with white shafts; lower back and upper tail-coverts, brownish black, margined with chestnut; central tail-feathers, dark brown, margined with chestnut; lateral feathers, pale brown, margined with white, all the tail-feathers lengthened and sharply pointed; sides of face and throat, white, the former spotted with brown; breast, dark buff; rest of under parts, white, with central brown streak on under tail-coverts; bill, black at tip, greenish yellow at base of mandible; feet and tarsi, greenish yellow. Length, 7½–9 inches; wing, 5–5½; culmen, 1; tarsus, 1⅛–1¼.

*Adult in Winter.*—"Above, grayish brown (more rusty on top of head), streaked and striped with dusky; superciliary stripe and lower parts, dull white; chest and sides of breast, pale grayish buff, the former indistinctly streaked with dusky; lower tail-coverts, with dusky shaft streaks."—*Ridgway.*

## PECTORAL SANDPIPER.

THIS species visits the United States during its migrations, appearing in April on its way northward, and again in August on its southern passage. It is met with all along the Atlantic Coast in small parties, and also in the interior. On the Pacific it is found in Alaska and in Puget Sound, but it does not seem to go southward by the way of the coast line of California, probably migrating inland to Central America, and so onward as far as Chili, by way of the shore. It is known by many names in the various parts of our country, some of which are: Krieker, Jack, Grass, and Meadow Snipe, Brown Bird, Short Neck, Hay Bird, Cow Snipe, etc. It has some game qualities, and will frequently lie to the dog, flushing like the true Snipe, and often fly off in a zigzag course, uttering loud, sharp cries, but it will not come to decoys. They frequent the salt meadows and muddy flats along the shores, also tidal creeks, and when feeding, if a number are present, they scatter all over the ground, busily searching for marine insects and small mollusks, occasionally thrusting the bill into the soil. At such times the birds are generally silent and not easily disturbed.

They run and walk with considerable swiftness, usually carrying the bill pointed slightly downward, and in search of food often wade into the shallow ponds, and if on the beach, follow the outgoing waves to pick up any insect or crustacean left by the waters. The Pectoral Sandpiper breeds in the far North,

20. Pectoral Sandpiper.

abundantly at Point Barrow, Alaska, according to Murdoch, in June and July, moving southward in September. Nelson, who has had most favorble opportunities for observing it during the love season, says it arrives near St. Michael's, on the shores of Behring Sea, about the middle of May, when the birds pair and build their nests. At this time they have a habit unique among waders, but often observed in members of the Grouse family, of inflating the throat until it becomes as large as the body, and uttering a note that is "hollow and resonant, but at the same time liquid and musical, and may be represented by a repetition of the syllables *tōō-ū, tōō-ū, tōō-ū.*" Nelson farther states that "the skin of the throat and breast becomes very flabby and loose at this season, and its inner surface is covered with small globular masses of fat." When not inflated it hangs down in a "pendulous flap or fold, exactly like a dewlap, about an inch and a half wide." "The bird may frequently be seen running along the ground close to the female, its enormous sac inflated and its head drawn back, and the bill pointing directly forward, or filled with springtime vigor, the bird flits with slow but energetic wing-strokes close to the ground, its head raised high over the shoulders and the tail hanging almost directly down. As it thus flies it utters a succession of the hollow booming notes, which have a strange ventriloquial quality. At times the male rises twenty or thirty yards in the air, and inflating its throat, glides down to the ground with its sac hanging below. Again he crosses back and forth in front of the female, puffing his breast out and bowing from side to side, running here and there as if intoxicated with passion. Whenever he pursues his love-making, his rather low but pervading note swells and dies in musical cadences, which form a striking part of the great bird chorus heard at this season in the North." The Pectoral Sandpiper is common in the delta of the Yukon, and the nest, usually placed in a tuft of grass, contains four eggs of a greenish drab color, spotted and blotched with umber brown, the average size being 1½ by 1.02 inches. In the autumn its habits in the North do not differ from those observed in its migration in southern climes.

### *TRINGA MACULATA.*

*Habitat.*—North America generally, south to Bermuda, West Indies, Brazil, and Chili in winter. Occurs frequently in Europe. Breeds in Arctic regions from Greenland to Alaska.

*Adult in Summer.*—Feathers of head and upper parts, together with scapulars and tertials, have brownish black centers, margined with pale buff, rufous, or white, giving a general pale appearance to the upper parts; wing-coverts, grayish brown with buffy-white margins; primaries, dark brown, shaft of first white; rump and upper tail-coverts, brownish black, margined narrowly with reddish buff; central tail-feathers, like upper coverts, lateral ones pale brown, margined with white; superciliary stripe, white, loral one dark brown; sides of head and throat, white, former with narrow brown streaks; breast and sides, pale buff, streaked with dark brown; rest of under parts, pure white; bill, apical half brownish black, basal half dull greenish yellow; legs and feet, buff. Length, 8½–9 inches; wing, 5–5⅘; culmen, 1–1¼; tarsus, 1–1⅕.

*Adult in Winter.*—Is without the rusty tints on the upper parts, and the black markings less sharply defined. The breast is also grayer.

## COOPER'S SANDPIPER.

MANY years ago this hypothetical species was described by Professor Baird from a single specimen shot at Raynor South, Long Island, by Mr. William Cooper. No second specimen has ever been obtained, and it is very doubtful if the type represents anything more than an unusually large White-rumped Sandpiper, or possibly a hybrid between that species and the Pectoral Sandpiper. The following description will assist any one to identify this bird should a specimen be obtained:

### *TRINGA COOPERI.*

*Habitat.*—One specimen taken at Raynor South, Long Island, nearly as large as the Knot, and almost exactly like the Pectoral Sandpiper. "Bill, straight, rather broad and a little widened at the tip, a little longer than the tarsus; tarsus, a little longer than the middle toe; hind toe and claw well developed; bare part of tibia a little more than half the tarsus, just half the bill; tail, doubly emarginate, but the central feathers projecting but slightly; upper parts, ashy gray, this being the color of the borders; the basal and central portion, however, is blackish, showing occasionally as a large spot. There are several scapular feathers which appear to be assuming a more perfect dress, and which are black, abruptly edged laterally with pale rusty, passing towards the tip into ashy. There is no rusty, however, on any other feathers. The head and neck are grayish, streaked with brown; the chin, whitish. The upper tail-coverts are white, each one with a V-shaped mark of brown; the rump feathers are brown, edged with whitish. The under parts are quite pure white, with a trace of reddish on the lower neck, but no indication of an ashy jugulum. The lower part of the neck, the jugulum, and the sides of the body show elongated oval spots of brown, not much crowded, but very well defined. These blotches under the wing are rather V-shaped, but where exposed are only in the end of the feather. There are also a few streaks in the crissum." Length, 9½ inches; wing, 5.75; tail, 2.80; tarsus, 1.14; middle toe and claw, 1 inch. —*Baird, l. c.*

## WHITE-RUMPED SANDPIPER.

BONAPARTE'S or Schinz's Sandpiper, by both of which names (in addition to the one given at the head of this article) this species is known, visits the eastern portion of North America on its migrations, not making a lengthy stay in any place. Coues observed it in Kansas migrating northward in flocks, and we may suppose it ranges from the Rocky Mountains eastward. Along the Atlantic Coast it appears at regular periods, passing northward in May, and back again on its southern journey in July or early in August. It associates with the Semipalmated Sandpiper (*E. pusillus*), which it somewhat resembles, but from which it is easily distinguished by its greater size. This species is one of the gentlest of all the waders, apparently paying little attention to an intruder upon its haunts, but allowing one to approach closely, not even suspending its occupation of searching for food. Should a gun be discharged as the little company draws itself together, the survivors fly a short distance in a compact flock, uttering a low, soft *tweet*, exhibiting the upper and then the under side of the body as they wheel and turn swiftly, and then frequently alight near the very spot where their companions were slaughtered. When on the wing it is recognizable by its white upper tail-coverts, which are very conspicuous. In Labrador it is very abundant, frequenting the rocky shores covered with seaweed or green and slippery from the flying spray. It also resorts to muddy flats and shallow pools, into which it

21. White-rumped Sandpiper.

wades up to the breast, in search of marine insects and various animalculæ, on which it feeds. It is a rather common bird at certain seasons on the shores of Lake Michigan, having been taken in Illinois, and also in Michigan. In the far north it is a straggler at Point Barrow in Alaska, and also breeds on the Mackenzie River. MacFarlane found the nest on the shore of the Arctic Sea, and on the Barren Grounds. This was merely a depression in the ground lined with a few decayed leaves, and contained three or four eggs, rufous drab in color, blotched with dark brown or black, confluent at the larger end, and measuring .35 inch long by .95 broad.

### *TRINGA FUSCICOLLIS.*

*Habitat.*—Alaska and eastern North America. Breeding in the Arctic regions from Greenland to Mackenzie River. In winter to the West Indies, South America, on the west coast, to Chili; Falkland Islands, where it is said to breed by Captain Abbott, who saw the young in East Falkland. (Dresser, H. E.) Occasional in the British Islands.

*Adult in Summer.*—Top of head and occiput, back and scapulars, black in the center of the feathers and margined with buff, or rusty buff, the latter prominent on back and scapulars, which also have the central black streak wide; wing-coverts, brown, margined with grayish white; primaries, dark brown; rump, blackish brown, feathers margined with gray; upper tail-coverts, white, streaked with dusky; middle tail-feathers, blackish brown; remainder, pale brown, edged with white; superciliary stripe, side of head, neck all around, breast and flanks, white, streaked narrowly with dark brown; throat and rest of under parts, pure white; some of the under tail-coverts with indistinct dark-brown streaks; bill, feet, and legs, greenish black. Length, 7½ inches; wing, 5; culmen, 1; tarsus, 1.

*Adult in Winter.*—Upper parts, brownish gray, indistinctly streaked with dusky; rest of plumage as in summer.

## BAIRD'S SANDPIPER.

BAIRD'S Sandpiper is only an occasional visitor to the Atlantic Coast, but is abundant in the interior, and in all suitable localities in the Rocky Mountains, large flocks having been seen in Colorado, at a height of about 14,000 feet, feeding on grasshoppers. It has usually been confounded with Bonaparte's Sandpiper (to which indeed it bears a close resemblance, but is rather smaller), and also with the Pectoral Sandpiper. It is about intermediate between the last-named wader and the Coast Sandpiper (*T. minutilla*), with a general similar pattern of coloration, but it has not such dark markings on the breast, nor the chestnut on the scapulars, as is seen in the last-named species. In its habits this pretty Sandpiper does not present any especial peculiarities differing from its allies; it associates with other waders, frequents the sandy margins of streams and muddy flats, is gentle, and has a low soft note. It is rather common at certain seasons in Minnesota, the Dakotas, Idaho, Montana, Colorado, Arizona, New Mexico, etc., keeping as a rule in small companies, large flocks being rather the exception. It also visits Illinois, but is not so common there as farther west. It seems to be more independent of wet lands and river banks than its relatives, and is found at times hunting for food at quite a distance from water. In Alaska, Baird's Sandpiper is not uncommon at Point Barrow, where it remains until after the breeding season. It occurs on the coast of Siberia, and has been

22. Baird's Sandpiper.

taken at St. Lawrence Bay; also at St. Michael's, and on Kadiak and Amak Islands, north of the Alaskan Peninsula. Dall states that it is not rare on the Yukon, and it has been procured at Sitka. This species arrives at Point Barrow in May before the snow has disappeared upon the plains, and keeps about the beach, retiring inland as the snow departs, and stays on the grassy portion of the plains. The nest, never on marshy ground, was hidden in grass, being merely a slight depression lined with grass. Mr. MacFarlane obtained the eggs on the Barren Ground in the Arctic regions, and here, contrary to what has been recorded by Murdoch at Point Barrow, the nest was placed in a swampy district, in a depression of the ground, carelessly formed of decayed leaves loosely laid and hidden by a tuft of grass. The female sits very closely, and only leaves the nest when the intruder is almost upon her, and then feigns lameness, and employs the other usual artifices to draw the observer away. The number of eggs is usually four, with a ground color of light drab uniformly sprinkled with spots and blotches of sepia brown. These markings are distributed generally over the whole shell, but are most numerous and nearly confluent at the larger end. They measured about 1 1/3 inches in length by 1 inch in breadth.

## *TRINGA BAIRDII.*

*Habitat.*—Interior of North America and western portions of South America, from the Arctic regions to Chili and Patagonia. Rare on the Atlantic Coast of North America and not found on the southern Pacific Coast of the United States. Breeding in the Arctic regions, in the valley of the Mackenzie River, and in Alaska.

*Adult in Summer.*—Top of head, back, and scapulars, variegated with black and grayish buff, the center of feathers being black, increasing in width on back and scapulars, and the margins grayish buff; wing-coverts, brown, margined with grayish white; primaries, dark brown, shaft of first one white; rump and upper tail-coverts, brownish black, margined with buff; side feathers of the coverts, white, with U-shaped black markings; middle tail-feathers, dark brown; remainder, pale grayish brown, margined with white; sides of head and breast, buff, streaked narrowly with dark brown; throat, abdomen, sides, and under tail-coverts, pure white; bill, feet, and tarsus, black. Length, 7½ inches; wing, 4¾; culmen, ¾–1; tarsus, about 1 inch.

*Adult in Winter.*—Does not differ much from the bird in summer plumage, the most noticeable variation being the feathers of the upper parts, which are margined with grayish white, giving a general gray hue to this part of the plumage; the head is more gray and lacks the dark buff seen in the summer specimens. In other respects there is no appreciable difference in the seasonal dress.

23. Least Sandpiper.

## LEAST SANDPIPER.

EVERYWHERE throughout the land where suitable localities are found, the *Peep*, congregated in flocks, is met with, migrating to the northward in May and reappearing in its accustomed haunts in July. On marshy meadows, the shores of creeks, rivers, and lakes, on fields of drifting seaweed, and, though not so often, on sandy beaches, the Least Sandpiper is one of the most abundant of our waders. Confiding and gentle in disposition, it betrays no fear of man, and goes quietly about its business picking up its minute articles of food (even when the observer stands within a few feet of it), uttering the while its low, sweet note. On taking wing, the birds mass closely together, fly swiftly, often in an an erratic course, showing alternately the upper and lower sides, and with shrill peeps circle round the object of their momentary alarm. Frequently the flock will alight close to the person who had intruded upon their domain, and its members immediately scatter about and commence feeding anew. They far outnumber the other species of waders, and wherever any of these are gathered together, there the little Peep will be found in the midst. On account of its small size it is not often hunted, the gunner, if he pays any attention to the birds at all, waiting until a number are massed together on the ground, and then makes a "pot shot" into their midst. At such times a great many are killed at one discharge of the gun. This species is abundant in Labrador, along the shores of Hudson's Bay, and north to the Arctic Circle. In Alaska it has been met with sparingly along the coast, but does not seem to visit any of the islands in Behring Sea. The nest is a depression in the dry moss, into which a little grass is placed, and generally four eggs are deposited, light drab or yellowish in color, spotted with dark brown, and, as is usually the case, most numerous at the larger end. They are pyriform in shape, and measure 1.15 by .85 inch.

The Least Sandpiper resembles very closely the Semipalmated Sandpiper (*Ereunetes pusillus*), which is equally common and widely distributed, but can always be distinguished from the last named species by having the toes completely cleft.

### *TRINGA MINUTILLA.*

*Habitat.*—All of North and South America, breeding from Canada northward to the Arctic regions, and from Labrador to Alaska. Occasional in Europe.

*Adult in Summer Plumage.*—Top of head, back, and scapulars, black, margined with reddish buff, the tips of feathers whitish; wing-coverts, grayish brown with pale edges, tertials edged with reddish buff; primaries, dark brown with white shafts; rump and upper tail-coverts, brownish black, edges paler, lateral tail-coverts with whitish bars on outer webs; middle tail-feathers, blackish with buff edges; sides of head, lores, neck, and breast, ashy buff, streaked with dark brown; superciliaries, throat, and rest of under parts, pure white, with indistinct spots of brown on throat; bill, legs, and feet, black. Length, about 5½–6 inches: wing, 3½–4; culmen, ¾; tarsus, ⅞.

*Adult in Winter.*—Plumage similar to that of summer, but more grayish and less bright, the breast ashy gray, indistinctly streaked with brown.

24. Long-toed Stint.

## LONG-TOED STINT.

THIS species is a miniature Sharp-tailed Sandpiper, without the lengthened tail and proportionately much longer toes. It also resembles somewhat the Least Sandpiper on the back, but exhibits very much more chestnut on the margin of the feathers of the back and head. The breast band is differently colored also. It is taken into the North American fauna from the fact that a specimen was procured on Otter Island, Alaska, June 8, 1885, as stated by Mr. Ridgway in the *Auk* for 1886. It is an Asiatic species, going as far east as Japan, and also found in the Indian Archipelago. It can only be regarded as an accidental straggler to our shores. On Behring Island, Stejneger observed it in large flocks in May, when it frequented the beach, and was very actively engaged picking up small crustaceans from the floating weeds which the surf had cast ashore. A few remained to breed, but the majority passed on farther north. He was unable to find the nest. The feet of this bird, when the legs are stretched out, extend far beyond the end of the middle tail-feathers.

*TRINGA DAMACENSIS.*

*Habitat.*—Asia, breeding northward, probably in the valley of the Lena, Otter Island, Behring Sea, Alaska.

*Adult, in May.*—Feathers of top of head and nape, black in center, margined with chestnut; back and scapulars, black, margined with ochraceous, chestnut, or grayish white; wing coverts, dark brown, margined with grayish; primaries, dark brown, first one with white shaft; rump and upper tail-coverts, black; lateral coverts, white, a few with central black streaks; middle tail-feathers, pointed, black; lateral feathers, pale brown, edged with white or buffy white; loral stripe, dark buff; superciliary stripe extending to nape, white; sides of head, buff; throat, white; band across breast, buff, indistinctly spotted with brown in the center, the spots larger, darker, and clearer on the sides, extending half way on the flanks; rest of under parts, pure white; bill, black; feet and legs, grayish yellow. Length, 5½ inches; wing, 3½; culmen, ¾; tarsus, ¾; middle toe, ¾.

## DUNLIN.

THE European Dunlin is only an accidental visitor to our shores. In pattern and color of plumage, there is no appreciable difference, save that the American bird is brighter in hue. In all the measurements, the Dunlin is rather smaller. It does not seem necessary to give a description of it, as that of the Red-backed Sandpiper will answer very well for both forms. The average measurements of the European bird are as follows: wing, 4.30–4.75 inches; culmen, 1.15–1.40; tarsus, .85–1; middle toe, .70–.75. The habits of the two forms do not differ.

*TRINGA ALPINA.*

*Habitat.*—Northern portions of Old World. In winter to North Africa, India, etc. Accidental in eastern North America (west side of Hudson Bay) and Long Island. Breeding up to north latitude, 74°.

## AMERICAN RED-BACKED SANDPIPER.

MOSTLY a bird of the seashore, the Red-backed Sandpiper frequents in large flocks the Atlantic Coast as far as Florida, where at times it is very abundant, and also that of the Pacific, at least as far south as San Francisco. It has been found in numbers in Kansas, Colorado, and the valley of Great Salt Lake. Beside the name already given, it is known by the gunners in various localities as the Black-bellied Sandpiper, Purre, Stib, Fall Snipe, Winter Snipe, Lead-back, Brant Snipe, etc. On our coast it frequents sand-bars, salt meadows, and muddy flats, arriving in April on its northern journey and returning in September. It is a very active little bird, continually moving about from place to place, very dexterous in obtaining its food of minute shellfish, worms, etc., the members of a flock keeping close together, both when flying and also when on the ground. The flight is very rapid, and the maneuvers on the wing are performed as if each and every individual was moved by one impulse. This Sandpiper migrates through the interior as well as by the seacoast, and proceeds in summer into the Arctic regions. In Alaska, where it breeds abundantly, it is found along the shore of Behring Sea north of Kotzebue Sound, is resident in summer at Point Barrow, common at the mouth of the Yukon and at Nulato, and in the south is found in Sitka, and on through British Columbia to California. When they arrive at the Yukon they are in full breeding dress, and the flocks

25. American Red-backed Sandpiper.

soon scatter, and pairing and nesting begin about the 1st of June. The males pursue the females, "uttering a musical trilling note which falls upon the ear like the mellow tinkle of large water drops falling rapidly into a partly filled vessel. It is not loud, but has a rich full tone difficult to describe, but pleasant to hear among the discordant notes of the various water fowl, whose hoarse cries arise on all sides. As the lover's suit approaches its end, the handsome suitor becomes exalted, and in his moments of excitement he rises fifteen or twenty yards, and hovering on tremulous wings over the object of his passion, pours forth a perfect gush of music, until he glides back to earth exhausted, but ready to repeat the effort a few minutes later." Murdoch says their rolling call is heard all over the tundra every day in June, and reminds one of the "notes of the frogs in New England in spring." The nest is usually placed on some slightly elevated dry knoll, though sometimes swampy places are selected, and on a bed of dead grass three or four eggs are deposited. In color they are pale greenish, or brownish clay, spotted and blotched with chocolate, and umber brown. Average size, 1.43 by 1 inch. Both sexes assist in the duty of incubation, and by the middle of July the young are able to fly, and about a month afterward they gather in flocks about the pools in company with the Pectoral Sandpiper and Long-billed Snipe, and by the last of the month all have departed for southern lands. In Davis Strait, Baffin's Bay on the eastern side of the continent, and on the islands of the Polar Sea this species, as stated by Sabine, was rare, but it was found abundant and breeding, by Ross, on Melville Peninsula and near Felix Harbor. Among individuals of the Red-backed Sandpiper there is often great variation in the color of plumage, and also in measurements, and many as small as the European are often taken on our own coasts. The birds from the two continents are very closely allied, and it is not always possible to separate them by means of any recognizable characters, and it is questionable whether anything is gained by attempting to divide them into races.

## *TRINGA ALPINA PACIFICA.*

*Habitat.*—Generally throughout North America, most numerous along the coasts. Breeding in the Arctic regions, perhaps not so far north as the European species. Eastern Asia, islands of Behring Sea.

*Adult in Summer.*—Crown and upper parts, except neck, including scapulars, bright rufous, streaked and spotted with black; wing-coverts, grayish brown, the greater bordered with white; primaries, pale brown; shafts, white; middle tail feathers, dark brown, remainder pale brown, lighter at the edges; sides of head, neck, and breast, grayish white, finely streaked with dusky; superciliary stripe and upper part of throat, white; abdomen, black; flanks, vent, and under tail-coverts, white streaked with brown; bill, feet, and legs, black. Length, 8½ inches; wing, 4½–5; culmen, 1½–1¾; tarsus, 1⅛; middle toe, ¾.

*Adult in Winter.*—Above ash gray, with brown shaft streaks, most conspicuous on the mantle and upper tail-coverts; breast, pale grayish white, indistinctly streaked with pale brown; rest of under parts, white.

26. Curlew Sandpiper.

## CURLEW SANDPIPER.

ALTHOUGH a very common species in the eastern hemisphere, we can only regard the Curlew Sandpiper as an accidental visitor to our shores. It has been taken a number of times on the Atlantic Coast, especially on Long Island, that former paradise of wading birds, and a few individuals in New England. In North Greenland it is stated to be not uncommon, and the eggs have been taken near Christianhaab. On the western coast in Alaska a male was procured by Murdoch at Point Barrow on June 6, 1883, and is the only instance of its appearance on our northwestern coast. It was in company with a flock of the Pectoral Sandpiper. In its habits the Curlew Sandpiper does not differ from the Red-backed Sandpiper, is an active little bird, and fond of associating with other species of waders. It runs upon the shore rapidly, carrying the head down, and it flies rather high and swiftly, with alternations of the back and breast as it wheels in its rapid course. Its food consists of small mollusks, crustaceans, worms, insects, etc., and it is said it also swallows the roots of marsh plants and small ground fruits, and feeds much at night. The eggs that were procured in Greenland by Governor Fencker and said to be of this species, were pyriform in shape, olive drab in color, blotched with two shades of umber brown, the markings most numerous at the larger end, and measured 1½ by 1.04 inches. The nest, which was but a slight hollow in the ground, lined with grass, was near the margin of a lake. Several times it has been supposed that the young in down and the eggs have been procured in the Old World, but in every instance it has been afterwards ascertained that they belonged to the Dunlin. I am not aware that any European or American naturalist has ever found the nest or eggs of this species, and Seebohm considers that no nest or eggs has ever been procured, alleging statements to the contrary to be myths. Farther evidence on this point is certainly desirable.

### *TRINGA FERRUGINEA.*

*Habitat.*—Northern portions of Old World, from the British Islands to China. In winter, to Africa, Indian Archipelago, and Australia. Accidental in eastern North America and Alaska. Breeding in Greenland and Arctic regions.

*Adult in Summer.*—Crown, hind-neck, back, and scapulars, black, margined with bright rusty; wing-coverts, brown, margins paler, the greater tipped with white; primaries, dark brown, shafts, white; rump, blackish brown; upper and under tail-coverts, white, spotted with black; tail, ashy gray, feathers edged with white; sides of head, neck, breast, flanks, abdomen and vent, rich chestnut rufous; some specimens have flanks and lower abdomen slightly marked with black; bill, legs, and feet, greenish black; iris, deep brown. Length, about 8½ inches; wing, 5; culmen, 1½; tarsus, 1⅛.

*Adult in Winter.*—Upper parts, brownish gray with indistinct streaks of dusky; rump, blackish gray, feathers margined with white; superciliary stripe, sides of head and throat, white, the last two streaked finely with gray; rest of under parts, white.

27. Spoon-bill Sandpiper.

## SPOON-BILL SANDPIPER.

I HAVE never seen this species alive. It is a native of eastern Asia, and has only twice been taken on our shores. Nelson obtained one in summer plumage in 1881 at Plover Bay on Choris Peninsula, and Dr. Bean the previous year had procured a specimen in autumn dress at the same place. Nelson says, "the bird was standing on the border of a small gravelly-edged pool on a spit at the entrance of the harbor. It had evidently fed to its satisfaction, and stood pensively at the water's edge when I came along. While watching it, before shooting, I saw it dabble its bill in the water and then draw in its neck, paying no further attention to its surroundings, although I was close by." In summer this species is found from Plover Bay to Cape Waukarun, and in this part of the Siberian coast it is probable that the breeding grounds are situated. At the last named place Nelson saw specimens of this Sandpiper in August. But few examples of this bird have been obtained. It is remarkable among the waders for the peculiar widening of the bill towards the point, from which it takes its trivial name, and this would at once attract the attention of any one who should capture a specimen. It is probably a rare species everywhere.

*EURYNORHYNCHUS PYGMÆUS.*

*Habitat.*—Asia. Arctic Coast in summer. In winter probably India or Burmah. Accidental in Alaska. Two specimens taken at Choris Peninsula. Breeding north of Behring Straits. Locality unknown.

*Adult in Summer.*—" Crown feathers, with blackish centers edged with rusty reddish, approaching chestnut; back of neck, with the dark centers becoming much fainter and the borders rufous, changing to buffy reddish, which, in addition to edging the feathers, appears to wash their surface and the dark central portions; the back and scapulars have well marked black centers, edged with rufous, buffy and grayish intermixed; the tertials have dark brownish centers, edged with grayish and russet; wing-coverts, light brown, edged with gray; the secondaries, largely white, and an imperfect wing bar formed by the white tips to the secondary coverts; quills, grayish brown, approaching black at the tips; the chin is whitish, washed with a pale shade of rufous, this latter shade becoming bright over the sides of the head and entire lower surface of neck, reaching the upper parts of the breast; the forehead, and around the base of the bill, washed with grayish over the rufous bases of feathers; the breast is rich buffy, changing to white on the posterior half of breast and entire abdomen; a scattered band of dark opaque shaft spots cross the breast and extend back on the sides, which are otherwise white; the tail is dark, ashy brown; bill, foot, and tarsus, black. Wing measures 3.95 inches; tail, approximately, 1.50; tarsus, .90; culmen, .90; width of expanded tip, .47. The hind toe is perfect but minute. The toes are not webbed."—*Nelson.*

*Adult in Winter.*—Forehead, cheeks, and all the under parts, white; upper parts, dusky, feathers margined with light gray.

*Young.*—" Back and scapulars, dusky, the feathers bordered terminally with dull whitish, these borders becoming rusty on anterior portion of back and scapulars; wing-coverts, dusky centrally, with still darker shaft streaks, and margined with brownish gray, the greater tipped with white; top of head, dull grayish, spotted with dusky, the feathers edged with rusty; rest of head, neck (except behind), and lower parts, white, clouded with light grayish brown and suffused with dull buffy anteriorly."—*Ridgway.*

28. Semipalmated Sandpiper.

## SEMIPALMATED SANDPIPER.

THE Peep, or Ox-eye, as it is often called, is one of the best known and most abundant of the Sandpipers, being seen everywhere on our eastern coast in places frequented by waders. It is also a migrant through the Mississippi valley, and at times fairly swarms upon the salt marshes, muddy flats, and banks of tidal creeks and rivers. It goes in large flocks, massing in close order when flying, uttering a soft whistling note, *peep-peep* or *tweet*, and on alighting scatters over the ground, and falls busily to work, probing the soil and picking up the minute creatures on which it feeds. At times the bird is very gentle and seems to pay no attention to the intruder on its domain, allowing a very near approach; but on other occasions it is more wary, ceasing to feed as soon as any one draws near. At such times all the members of a flock stand motionless for a few moments, and then rise simultaneously with shrill cries, drawing together into a compact mass, and with many evolutions and rapid twistings, fly around and away for a short distance ; then frequently returning (especially if an imitation of their cry is given) near to the spot from which they started, when they alight and remain motionless, not scattering as usual to feed. If the object that alarmed them is still present, after watching a few moments they rise again and generally seek some other feeding ground. It is met with in Texas, also in British Columbia, and it is not improbable that flocks of the representative race in the west, the Western Sandpiper (*E. occidentalis*), often contain individuals of this species. To one not looking for them, or apt to notice the slight difference existing, the two forms would easily be confounded, and the presence of either in an unlikely locality pass without remark. It is said to be fairly abundant at Point Barrow, Alaska, and along the coast of the Alaskan Peninsula, and in some of the islands of Behring Sea. At Franklin Bay in the Barren Grounds, and on the islands in the Arctic Sea, this Sandpiper has been found breeding. The nest was a depression in the midst of some dried grass, and the eggs, usually four in number, pyriform in shape. The color varies greatly, from a drab to a light grayish buff. The spots, sometimes large and distinct, again minute and very numerous, hiding the ground color, are reddish or dark sepia in hue. They measure 1.15 to 1.25 inches in length, by .80–.90 in breadth.

## *EREUNETES PUSILLUS.*

*Habitat.*—Eastern North America, south in winter to the West Indies and South America. Breeds in the north to the shores of the Arctic Sea, and from Labrador to Rocky Mountains.

*Adult in Summer.*—Upper parts with head, grayish brown; central portion of feathers on back and scapulars, black, margined with pale buff; wing-coverts, pale brown, margined with whitish; primaries, dark brown, inclined to black at their tips; rump, black, feathers margined with brown · upper tail-coverts, blackish brown; side-coverts, white, with brown streaks; middle tail-feathers, dark brown, remainder grayish brown; dusky loral stripe; ear-coverts, light brown, streaked with dusky; superciliary stripe and throat, pure white; breast, ash, striped with brown; rest of under parts, pure white; bill, feet, and legs, black. Length, 5¼ inches to 6½; wing, 3¾; culmen, ¾; tarsus, ⅞.

*Adult in Winter.*—There is not very much difference in the winter dress on the upper parts, which are perhaps more brownish, but the breast is merely washed with grayish, and the streaks are very indistinct, sometimes obsolete.

29. Western Semipalmated Sandpiper.

## WESTERN SEMIPALMATED SANDPIPER.

THIS is the western form of the Semipalmated Sandpiper, being very abundant along the Pacific Coast, and also in Arizona and Texas as a migrant in May and September. Occasionally individuals are met with on the Atlantic Coast in company with the Least Sandpiper, but it can only be regarded as a straggler in the eastern sections of the Union. It is much more rusty or chestnut on the upper parts than the Semipalmated Sandpiper, and the bill has a considerably greater average length. Nelson found it abundant about Norton Sound and on the shore of Behring Sea, near St. Michael's, and the mouth of the Yukon, where it arrived about the middle of May. Towards the end of the month they forsake the borders of the pools and scatter over the tundra, and the male begins his courting of the shy, retiring female, who "modestly avoids the male as he pays his homage, running back and forth before her as though anxious to exhibit his tiny form to the best advantage. At times his heart beats high with pride as he trails his wings, elevates and partly spreads his tail, and struts in front of his lady fair in all the pompous vanity of a pigmy turkey-cock ; or his blood courses in a fiery stream until, filled with ecstatic joy, the sanguine lover springs from the earth, and rising upon vibrating wings some ten or fifteen yards, he poises, hovering in the same position sometimes nearly a minute, while he pours forth a rapid, uniform series of rather musical trills, which vary in strength as they gradually rise and fall, producing pleasant cadences. The wings of the songster meanwhile vibrate with such rapid motion, that they appear to keep time with the rapidly trilling notes, which can only be likened to the running down of a small spring and may be represented by the syllables *tzr-r-e-e-e zr-e-e-e, zr-e-e-e*, in a fine pitched tone with an impetus at each *z*. This part of the song ended, the bird raises its wings above its back, thus forming a V and glides slowly to the ground, uttering at the same time in a trill, but with a deeper and richer tone, a series of notes which may be likened to the syllables *tzur-r-r-r, tzur-r-r-r*." The nest is usually placed on a mossy hummock on slightly elevated ground, sheltered by grassy stems, the male remaining in the vicinity while the female is on the nest. The female when disturbed feigns lameness, or utters low, plaintive notes as if beseeching the intruder to spare her treasures. The eggs are a pale clay color, shading to brownish clay, covered with fine light reddish brown spots, or chocolate blotches, most numerous at the larger end. The old and young leave in September for the south, the last loiterer having gone by the 1st of October. In its habits this species does not vary from the Least Sandpiper, and is of an equally gentle and trusting disposition.

## *EREUNETES OCCIDENTALIS.*

*Habitat.*—Western North America, from Alaska south in winter to Central and South America. Occasional on the Atlantic Coast. Breeding in Alaska.

*Adult in Summer.*—Feathers of crown, nape, back, and scapulars, centrally blotched with black and margined with rusty and grayish white; the rust color sometimes is uniform on sides of crown and neck; wing-coverts and primaries, like *E. pusillus*; rump and middle tail-coverts, brownish black, slightly edged with pale brown or rufous; lateral tail-coverts, white, some with brown streaks; tail, like *E. pusillus;* loral stripe and ear-coverts, rusty, finely streaked with brown; throat, sides of neck, and under parts, pure white, the neck and breast thickly streaked with blackish brown; bill, feet, and legs, black; wing, average 3¼ inches; culmen, .95; tarsus, ⅞.

*Adult in Winter*—This species can only be distinguished in the winter dress from *E. pusillus* by the greater length of bill and tarsus. Some specimens may have more rusty on the upper parts, but the general appearance is very much the same.

## SANDERLING.

SANDERLING, Beach Bird, or Ruddy Plover, by all of which names this species is known, is distributed throughout the globe, and is found on both coasts of North America as well as in the interior, and is one of the most familiar of our waders. It breeds in the far north, but seems to be rather rare in portions of Alaska, although Dall found it common at Nulato, and on the Yukon to the sea. It usually frequents the beaches, running over the sand with great rapidity, keeping close to the margin of the water and following the receding waves, picking up with great dexterity the minute molluscæ and insects left by the retiring surf. Its agility in avoiding the incoming billows and rush of the seething foam-covered water as it mounts the beach is extraordinary, and often it wades into wavelets up to its belly to seize some coveted morsel. It is generally rather wary, not permitting a very near approach when so occupied, the little flock if disturbed rising and with shrill whistles flying for a hundred yards along the beach, then alighting and commencing again to feed. When flying individuals keep closely together, the wings curved downwards, and moved with rapid, jerky beats.

This species is the commonest of the beach birds on our shores, and can almost always be found along the margin of the water during the season when any of the waders are present within our limits. It may occasionally be seen upon the shoals and mud flats near bays

30. Sanderling.

and pools, but it is essentially a beach bird, preferring the vicinity of the ocean, over which it frequently makes long flights at some distance from the land. The Sanderling is generally seen in moderately sized flocks, but many such may be observed at various distances apart, along a vast extent of sea-beach, and each little company keeps pretty well to itself, not often mixing with others, except when the gunner has reduced its ranks by several discharges of his gun, when the survivors will usually unite with some other flock that has been more fortunate in keeping out of danger. In the spring and autumn migrations it is not uncommon about the Great Lakes, and has been observed in various States in the Mississippi Valley, but it is most abundant on the seacoasts. When running over the sands it has the habit of raising its wings above the back, as if just about to fly, and if wounded will often take to the water, upon which it floats lightly and swims easily. Its foods consists of small crustaceans, worms, insects, and in the far north it has been observed to eat the buds of the saxifrage. Not much is known of the breeding habits of this interesting little bird. MacFarlane found a nest in the Barren Ground, east of Anderson River, close to the Arctic Sea, and Captain Fielden observed it breeding in Grinnell Land, west of Cape Union, on the shores of the Frozen Ocean. The Barren Ground nest was composed of hay and decayed leaves, and contained four eggs, and the female was secured. The nest found by Captain Fielden was merely a depression in a plant of willow lined with leaves and catkins. The male was killed, which seems to show that both sexes incubate the eggs. These are buffish or brownish olive, faintly spotted with olive brown or bister, with underlying markings of olive gray. Sometimes the markings are mostly at the larger end, but again they are pretty evenly distributed over the surface.

### *CALIDRIS ARENARIA.*

*Habitat.*—Cosmopolitan. Breeding in Arctic regions, Anderson River, Grinnell Land, Greenland, Sabine Island, Iceland. South in winter to Chili and Patagonia.

*Adult in Summer.*—Crown, nape, mantle, and scapulars have the center of all the feathers black, margined with rufous and grayish white, the latter hue very conspicuous; wing-coverts, ashy brown, margined with rufous or grayish white; the greater coverts, margined with white, forming quite a broad bar across the wing; primaries, dark brown, blackish towards the tips; shafts, white; rump, dark brown, feathers margined with grayish white, mixed with rufous; middle tail-covets, black, margined with rufous and whitish; lateral tail-coverts, white, with occasional black central streaks; tail, grayish brown, central pair darkest; side of head, throat, neck, and breast, light rufous, speckled and streaked with black; rest of under parts, pure white; bill, legs, and feet, jet black. Length, 8 inches; wing, 5; culmen, 1; tarsus, 1.

*Adult in Winter.*—Upper parts, pale gray, center of feathers black; rump, pearl gray, center of feathers light brown; entire under parts, pure white; tail as in summer, perhaps a little lighter.

31. Marbled Godwit.

## MARBLED GODWIT.

INHABITING North America generally, the Marbled Godwit breeds in the Mississippi valley, also in the Missouri region, and from the Saskatchewan to Texas, and as far east as Western Pennsylvania, as I am informed by Mr. George B. Sennett, going in winter to the Argentine Republic. In the Arctic regions it is at times very abundant, and also on the Saskatchewan plains; but it has not been obtained in Alaska. It arrives in May on the Atlantic Coast of the United States on its northern migration, in moderate sized flocks, but cannot be said to be anywhere very numerous. It frequents the salt marshes and borders of pools and ponds, is very watchful and wary, and does not permit a near approach, taking wing soon after it has noticed an intruder on its domain. Like the Curlews and its relatives in the same genus, members of a flock are greatly attached to each other, and if any are wounded the others will return again and again to the place, and hover over their luckless companions, frequently meeting the same fate that has befallen them. This species is known in various localities by many names, some of which are: Red Curlew, Straight-billed Curlew, Marlin, Horsefoot Marlin, etc. On the Pacific Coast this species is found in numbers in Southern California, and probably breeds there, as Cooper states that the young make their appearance in May near San Pedro. Sometimes this Godwit goes in immense flocks, like that mentioned by Audubon, near Cape Sable, in Florida, when he saw thousands collected on a mud bar; but usually there are not more than twenty or thirty observed together. Like all the waders, however, they are met with on our eastern coast yearly in diminished numbers; improved firearms and the constantly increasing army of gunners serving to reduce their ranks, until they have been entirely driven away from many localities where formerly they were abundant. The nest is usually placed near the water, and is a depression in the ground lined with grass; the color of the egg is an olive drab, spotted with various shades of yellow and umber brown. They measure 2.27 by 1.60 inches.

## *LIMOSA FEDOA.*

*Habitat.*—North America generally. Breeding in the interior, from Iowa and Dakota, north to Lake Winnipeg. Southward in winter to Cuba, Guatemala, Yucatan, and South America to Argentine Republic on the east coast.

*Adult.*—Head and neck, pale buff, streaked with black, the streaks broadest and most numerous on the crown and nape; entire upper parts and scapulars, reddish buff, irregularly barred with black, sometimes the bars becoming confluent; wing-coverts, reddish-buff, barred with dusky; first three primaries, dark brown on the outer web and tips, reddish brown on the inner, speckled with black; remaining primaries, reddish brown, speckled with blackish brown on both webs, and with a subterminal blackish-brown bar; rump and upper tail-coverts, buff, diagonally barred with dark brown; tail, reddish, irregularly barred with dark brown; broad, white stripe from bill to above the eye; loral stripe, dark brown; throat, white; entire under parts, pale rufous, varying in intensity among individuals, transversely crossed with wavy dark-brown lines on all the feathers, save the center of abdomen and vent, which is a uniform pale buff; under wing-coverts, reddish buff; bill, flesh color on basal half, blackish brown on the remaining part; feet, dark bluish gray. Length, 16½–20½ inches; wing, 8¾; culmen, 3¾; tarsus, 2¾; middle toe, 1½.

32. Pacific Godwit.

## PACIFIC GODWIT.

A NATIVE of eastern Asia, the Pacific Godwit is a summer resident of Alaska, and also an occasional visitor to Southern California, having been taken by Mr. Belden at La Paz, as recorded by Bryant. In Alaska, Dall found it very plentiful about the mouth of the Yukon, and it is sometimes seen at Point Barrow. At St. Michael's, Nelson saw them in flocks of from twenty-five to two hundred. They were shy, constantly in motion, wheeling and circling over the land, alighting occasionally, but only for a moment. By the middle of May the courtship begins, the flocks break up, and the birds scatter over their breeding ground. The males, he says, utter a loud ringing *kŭ-wēw*, *kŭ-wēw*, with great emphasis on the last syllable, which can be heard for several hundred yards. They frequent the grassy open country, and should any one enter their territory they protest against the intrusion by circling around and uttering ear-piercing cries of *kŭ-wēw*. If the nest or young is near they swoop close to the offender's head, redoubling their cries. The same note is heard during the courtship, and they also have a rolling whistle at this season resembling that of the field Plover. When flying the wings are decurved, and with a few rapid strokes they sail for a short distance, then repeat the beats. It walks well and gracefully with head well up, and frequently raises the wings high above the back and then folds them again. Like most of their tribe this species comes readily to decoys if their note is well imitated. The young are on the wing by the middle of July, and by the end of August all have departed. Nelson says that in the breeding season all the birds in the neighborhood will unite to escort a dog through their territory with the most resounding cries. This species is a regular visitor to the Commander Islands during the migratory season, but does not seem to breed there. The large size of this bird and its loud voice makes it one of the most conspicuous of the waders in the countries it frequents. Dall found it very common on the Pastolik marshes north of the Yukon mouth, as well as the place where the great river enters the sea, and he says it is the largest snipe found in the country. The nest is a rounded depression in a sedge tussock lined with grass, and the eggs, two in number, are light olivaceous in color, spotted with various shades of brown, resembling somewhat those of the Marbled Godwit. They measure 2.25 inches in length by 1.45 in breadth.

## *LIMOSA LAPPONICA BAUERI.*

*Habitat.*—Shores and islands of the Pacific Ocean from Australia to Alaska. On the American coast, Alaska in summer; and also from La Paz, Lower California. Breeding possibly in Siberia.

*Adult in Summer.*—Top of head and hind-neck, streaked with blackish brown, feathers edged with yellowish white; back and scapulars, blackish brown, feathers margined gray, buff, or rufous; wing-coverts, ashy brown, bordered with whitish, the lesser coverts darker than the greater; primaries, brownish black on outer web, paler on inner, and with white shafts; rump, dark brown, bordered with white; upper tail-coverts, white, barred with dark brown, some of the central ones cinnamon; tail, brownish gray, barred irregularly at base, and for two-thirds the length of inner webs with white, and tipped with white; lores, dark brown; stripe from bill to eye, buffy white; entire under surface, buffy red, barred irregularly on flanks and under tail-coverts with dark brown; bill, flesh color on basal half, blackish brown for the remainder. Length, 16 inches; wing, 8¾; culmen, 3½; tarsus, 2¼.

*The Female in Summer.*—Like the male, but the under surface is paler and mixed with white, as if immature, and she averages larger in all measurements.

*Adult in Winter.*—Crown, back of neck, and upper parts, brownish gray, lightest on head and neck, with the center of feathers dark brown; under parts, brownish ash on throat and neck, white on breast and abdomen, faintly barred on sides. There seems to be a wide variation in color and pattern among individuals in their plumage.

## HUDSONIAN GODWIT.

THE Hudsonian Godwit, or Ring-tail Marlin, is, according to my experience, more often met with, especially on Long Island and the New Jersey Coast, than the Marbled Godwit, but is not common at any time. It is found generally throughout the country east of the Rocky Mountains, but never, so far as I am aware, on the Pacific Coast of the United States south of Alaska. It is a regular visitant, during migration, to the States bordering on the Great Lakes, and is often procured in full breeding plumage in Minnesota, Dakota, and some other of the adjacent States. It frequents the marshes and salt meadows near the sea, and from its large size is very conspicuous among the other waders scattered about in its vicinity. It is known by many names to the gunners, such as Red-breasted and Rose-breasted Godwit, Goose-bird, Black-tail, White Rump, Carolina Willet, etc., but perhaps that of Ring-tail Marlin, to distinguish it from its relative, is the most familiar. Its white upper tail-coverts are clearly shown when the bird is flying, contrasting strongly with the dark rump and tail. It is rare on the Atlantic Coast in the full breeding dress, examples taken being in the young plumage, or in partial summer costume. Like the other Godwit, its larger relative, it is a shy bird during migration, and keeps a watchful eye on an intruder on its domain, rising at a considerable distance and uttering its shrill cry. It sometimes decoys readily, setting

33. Hudsonian Godwit.

its wings and sailing up to the wooden counterfeits, lured on by a close imitation of its note ; but soon discovers the deception and either alights only for a moment, or else wheels about over the decoys, and hastily departs, provided it escapes the rain of shot from the discharged gun of the concealed sportsman. About Hudson's Bay it is met with in large flocks, resorting to the beach when the tide is low, and feeding on the crustacea it discovers there, retiring to the marshes as the tide rises.

It has been taken at Great Slave Lake, and on the Anderson River, and MacFarlane found it breeding in the vicinity of Fort Anderson. It also breeds on the Barren Grounds near the Arctic Ocean. The nest, formed of decayed leaves, is placed in a depression in the ground, and the eggs, usually four in number, pyriform in shape, are dark olive drab, spotted and blotched with reddish brown, measuring 2.15 to 2.22 in length by 1.39 to 1.41 in breadth. In Alaska the Hudsonian Godwit appears to be rare ; Dall obtained two specimens at the mouth of the Yukon, and one was procured at Nulato. Nelson says it occurs more commonly at Fort Yukon as a migrant, but although it undoubtedly breeds in the Territory, its nest has not been discovered. In the Argentine Republic, where many individuals of this species pass our winter, the summer of the southern hemisphere, they appear on the pampas in April and remain until September. They associate in small flocks of from twelve to thirty individuals.

Birds from both the Arctic and Antarctic regions must meet and mingle together, a curious circumstance, remarks Hudson, "so far from the breeding place of one set of individuals and so near to that of the other." In May the species has been observed in the Falkland Islands.

### *LIMOSA HÆMASTICA.*

*Habitat.*—North America, east of Rocky Mountains; also in Alaska; migrating in winter to Cuba, and in South America to the Argentine Republic and to Chili. Breeding from Baffin's Bay to Alaska.

*Adult in Summer.*—Head, back, and sides of neck, grayish white, sometimes suffused with buff, and streaked narrowly with black; back and scapulars, black, feathers with spots or bars of buff and edged with grayish white, this edging in some examples almost obsolete, when the buff spots become more conspicuous; lesser wing-coverts, dark brown, margined with pale brown; greater coverts, dark brown, margined with white, forming an obscure bar across the wing; primaries, blackish brown, shafts white; secondaries, blackish brown, with buff and white spots or broken bars on outer webs; axillaries, brownish black; rump, brownish black; upper tail-coverts, pure white; tail, black, the base and tip white; superciliary stripe, white, speckled with dark brown; throat, buffy white, narrowly streaked with black; lower parts, dark chestnut, narrowly barred with black; feathers, margined with grayish white on the sides and on the abdomen; under tail-coverts, white, barred with black, sometimes with buff also; bill, flesh color, black for apical third; feet and legs, grayish blue; iris, brown. Length, 14–16 inches; wing, 8; tail, 3½; culmen, 3; tarsus, 2⅛.

*Adult in Winter.*—Head, neck, and lower parts, grayish buff, shaded with brownish gray; upper parts, dark grayish brown; wings, rump, tail, etc., as in summer. The young resemble the winter plumage, but the feathers of the back have a subterminal blackish-brown bar, edged with buff; beneath, pale buffy gray.

34. Black-tailed Godwit.

## BLACK-TAILED GODWIT.

AN occasional appearance in Greenland is the only claim this species has to a place in the North American fauna. It belongs to the Old World, not common in Great Britain, but breeds in the northern part of the continent of Europe, as far westward as the coast opposite the British islands. It migrates from its African winter quarters in large flocks and spreads over various portions of the European continent. It breeds in Poland, making a depression in a tussock and lines it with grass. Four eggs are deposited, dull greenish in color, marked with dark brownish olive, and the birds resent with loud cries any intrusion into their domain, as has been already mentioned in the description of the Pacific Godwit. In its habits this species does not seem to differ from its allies of this genus. It is said to fearlessly attack any cow or horse, or pursue any hawk or crow that may approach its nest. It occurs in India, China, and North Australia, and represents our Hudsonian Godwit in the Old World.

*LIMOSA LIMOSA.*

*Habitat.*—Northern portions of Old World. Accidental in Greenland. Breeding in northern Europe and in Arctic regions of old World to 64° north latitude.

*Adult in Summer.*—Head and neck all around, dark cinnamon; crown and nape, streaked with black; back and scapulars, black, margined with grayish brown and ferruginous; wing-coverts, brownish gray, the greater tipped broadly with white, forming a bar on the wing; secondaries, white on outer web; primaries, brownish black, pale on inner web, with white shafts; under wing-coverts, white; rump, very dark brown; upper tail-coverts, pure white; tail, black, basal half white, center feathers tipped with white; breast, pale cinnamon, barred with dusky; rest of lower parts, white, barred with black and cinnamon; middle of abdomen and under tail-coverts, nearly pure white, with a few brown bars; bill, blackish brown, orange at base; legs and feet, black; iris, brown. Length, 15¾ inches; wing, 8; culmen, 3¾; tarsus, 2⅔.

*Adult in Winter.*—Wings and their coverts, lower back, upper tail-coverts, and tail, same as in summer; head, neck, upper back, and scapulars, dark brownish gray: fore-neck, pale gray; lower parts, white.

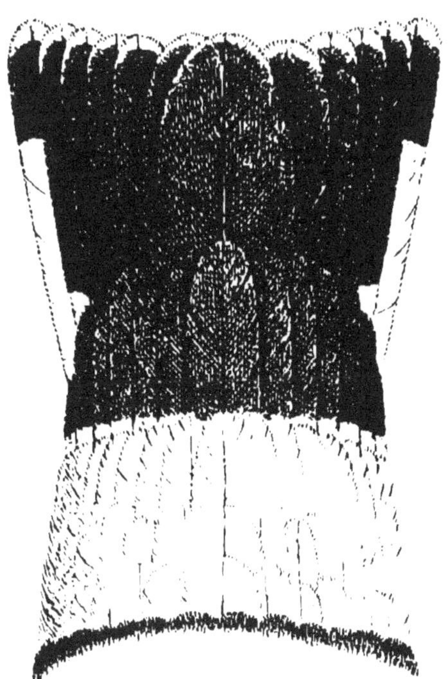

35. Tail of the Black-tailed Godwit.

36. Greater Yellow-legs. 37. Little Yellow-legs.

## GREATER YELLOW-LEGS.

ALTHOUGH breeding within the limits of the Union, the Big Yellow-leg is known throughout the United States mostly as a migrant. It appears along the Atlantic Coast in the spring, about the middle of May on Long Island, on its way to its northern breeding grounds, returning again in August. As a rule it does not make a very protracted stay in any locality, the earlier arrivals making way for those succeeding, and by the beginning or middle of October the last straggler has moved on towards its winter home in the South. It is a common species throughout the interior wherever water is plentiful, and is numerous on the coast of California, where it may be found about the marshes, both fresh and salt, nearly throughout the year. It may possibly breed in that State. This bird is known by many names in the different localities it frequents, some of which are Telltale Snipe, Yelper, Winter Yellow-leg, Yellow-shins, Cu-cu, and Large Cu-cu, Stone Bird, etc. It is a noisy species, continually uttering its shrill whistling note, by imitating which it can be easily brought to the decoys, even when the flock is flying at a considerable elevation. When the birds first hear the familiar call, they check their onward progress, and on stationary wings sail around the spot in a wide circle, as if trying to ascertain the exact locality from which the sound proceeds. Then, as the three notes composing the call are repeated, they catch sight of the decoys, and rapidly descending, sometimes in a zigzag, erratic course, they fly to leeward of their counterfeit representatives, and turning against the wind, with shrill whistling cries, sail up in compact ranks and hover for a moment over the wooden images, preparing to alight in their midst. Then is the sportsman's opportunity, who, rising in his blind of grass or reeds, discharges both barrels of his gun into the midst of the jostling birds, strewing the ground with the dead and wounded. The survivors, with cries of alarm, on quickly moving pinions hurry from the dangerous spot, and when but a little distance away the familiar whistle again falls upon their ears, and circling round with trustful confidence they return to the place where their comrades are lying. Once more the fatal storm of shot bursts upon them, thinning their ranks, when, now fairly alarmed, the few remaining birds depart swiftly, paying no further heed to the hidden charmer, charm he ever so wisely and well. This species on alighting raises its wings high over the back, as if, distrusting the ground, it was about to take flight, and then slowly folds them into their proper place. Sometimes on the Atlantic Coast there are great flights of this snipe, and again it is rather scarce, at all events in comparison with other waders. I do not think it is ever so numerous as the succeeding species. It keeps to the marshes and shores of tidal creeks and rivers, also the muddy flats that are often spread out to a great extent, feeding on small crustacea, insects, worms, etc., and often wades into the water up to its belly. It walks easily and with somewhat of a stately carriage, holding the head well up and keeping a sharp lookout for all intruders. The eggs of the Big Yellow-leg have not been often taken. They are grayish white, spotted with various shades of brown and lilac covering the surface, most numerous at the larger end. The nest is but a slight depression in the ground near the water and lined sparsely with grass. The eggs are four in number, and measure 1.70 by 1.30 inches. In South America the Big Yellow-leg arrives in the Argentine Republic in October, and is quite common about the ponds and marshes. It leaves for the north in March, and its place is then supplied by others arriving probably from the Antarctic regions, coming north to the pampas to pass the winter. So the species is found in that country all the year round, its supplies having been drawn from two independent sources. It is not known, however, to breed there, the Antarctic birds going farther south for the purpose of incubation, the others proceeding to the northern portions of North America for the nesting season.

### *TOTANUS MELANOLEUCUS.*

*Habitat.*—North America generally, migrating in winter to Argentine Republic and Chili. Breeds from the Northern States, such as Illinois, Wisconsin, Iowa, etc., northward. Found by Audubon breeding in Labrador. Probably also in Alaska, although its nest has not yet been found there.

*Adult in Summer.*—Head and neck all around, white, streaked with black; back and wings, variegated with black and gray and white, the markings in the shape of triangular spots at the edge of the webs; primaries, brownish black; rump, brownish black, feathers edged with grayish white; upper tail-coverts, white, irregularly and narrowly barred with black; middle tail-feathers, pale gray, barred with dusky; remainder, white, barred with black; throat and entire under parts, white, streaked conspicuously with black on breast, and barred with same on flanks and under tail-coverts; bill, black; feet and legs, yellow. Length, 15 inches; wing, 8; culmen, 2¼; tarsus, 2½; middle toe, 1½.

*Adult in Winter.*—Similar to the summer dress, but the upper parts and wings are ash gray, with little or no black, and the margins of the feathers and spots white; under parts, pure white, the breast and flanks faintly marked with brownish gray.

## LITTLE YELLOW-LEGS.

THE Little Yellow-legs is the Big Yellow-legs in miniature, the plumage being almost exactly the same in pattern and coloration. It is more numerous than its large relative, and is found generally throughout the United States, but not so abundant west of the Rocky Mountains, except perhaps in Alaska, where it has been found from Sitka to the Yukon, on the upper portion of which river it is common. It has also been obtained on some of the islands in Behring Sea. This snipe is very sociable, goes in flocks of considerable size, and is always calling for others to come and join it. Its cry is very similar to that of the larger species, consisting of three shrill notes rapidly uttered; and the habits do not vary from those of its relative. It is easily decoyed, more so than the Greater Yellow-legs, and as it approaches the lures lowers its long legs and hovers over them. On the seacoast, like all the waders, it is apt to have a sedgy flavor, but this is not apparent in the birds obtained in the interior away from salt water. This bird is known as Summer Yellow-leg, Tell-tale, and by many of the names bestowed on the other species. It breeds in the Arctic regions, probably across the continent in the same high latitudes. Its nests have been found at Great Slave Lake and in the Anderson River district. It resents intrusion on its breeding grounds, flying about with hanging legs and drooping wings, and uttering incessantly loud shrill cries. While of gentle disposition, it is very watchful, and often alarms other waders and causes them to take wing, much to the chagrin of the sportsman who has waited patiently for a long time in his cramped quarters and uncomfortable position in hopes he may draw them by well-imitated notes to his place of concealment. This species visits the shores of Hudson's Bay in large flocks on its southern migration. It passes regularly through the valley of the Mississippi in the spring and autumn, associating often with the Big Yellow-leg, and is usually most abundant when going southward. It breeds in Illinois and Ohio, possibly farther south, as well as in the far north. The nest is but a depression in the ground, placed amid the grass, near water, lined with leaves or twigs. The eggs, generally four in number, are of a light drab color or brown, blotched with chocolate or rufous, sometimes with a much paler tint, pyriform in shape, and measure 1½–1¾ inches in length by 1⅙ in breadth.

### *TOTANUS FLAVIPES.*

*Habitat.*—North America, less common in the west than in the eastern provinces, going in winter to Patagonia. Breeding in the Arctic regions from the Yukon to Greenland; is not rare in Alaska, and is found in Greenland.

*Adult in Summer.*—Crown and nape, mixed brown, black, and white, the latter hue on the edge of the feathers; neck, white, streaked with dark brown; back and scapulars, ashy brown, with blotches of black and spots of white; primaries, dark brown; rump, brownish black, feathers edged with white; upper tail-coverts, pure white, with occasional black bars, some indistinct; central tail-feathers, gray, barred with brown and white; outer feathers white, barred with brownish black; under parts, pure white, streaked with black on neck and breast, and barred with same on flanks and some of the under tail-coverts; bill, black; feet and legs, yellow. Length, 11 inches; wing, 6¼; culmen, 1½; tarsus, 2⅛; middle toe, 1.

*Adult in Winter.*—Upper parts, ash, variegated on scapulars and back with white spots; head, neck, and breast, ash, with white streaks; rest of under parts, pure white; other parts, as in summer.

## EUROPEAN GREENSHANK.

AUDUBON obtained three specimens of this bird on Sand Key, near Cape Sable, Florida, and this is the only recorded instance of its occurrence in North America. It is a native of the Old World, and is common in many portions of the continent of Europe during the migrations; also is distributed generally through China in winter, and in the island of Hainan it goes in flocks of thousands. It is common in India, and in Africa is found from the Mediterranean to the Cape of Good Hope. In the British islands the Greenshank is a rather wary bird, and can be approached when feeding only with difficulty. It is said to resemble the Avocet in the manner of procuring its food, placing the bill upon the surface of the water and swinging it from side to side, leaving a zigzag line traced upon the mud at the bottom. This species swallows water insects and their larvæ, small beetles, tadpoles, and sometimes fish and frog spawn, worms and small fish. The nest is a depression in the grass in marshy ground, and the eggs are buffy white, spotted with dark brown, with underlying markings of purplish brown.

### *TOTANUS LITTOREUS.*

*Habitat.*—Old World. Accidental in Florida. Breeds in Siberia from latitude 60° to 66°.

*Adult in Spring Plumage.*—Head, neck, scapulars and back, striped with black on a gray ground, and with the margin of feathers white; wing-coverts, dark brown, some edged with white; primaries, blackish brown,

38. European Greenshank.

shaft of first white; lower back, rump, and upper tail-coverts, white, the latter barred irregularly with black; central tail-feathers gray, barred with dusky, remainder pure white, the two next the central pair barred with black; entire under surface, white, neck and breast spotted with black; bill, black; legs and feet, olive green. Length, 14½ inches; culmen, 2¼; wing, 7½; tarsus, 2¼.

*Adult in Winter.*—Above, pale ash gray, the feathers edged with white; forehead, white; lores and center of forehead, dusky black; under surface, pure white.

## SOLITARY SANDPIPER.

THE Solitary Sandpiper, or Wood Tattler, is a migratory species in the United States. It breeds within the limits of the Union and also farther north, but passes onward, as the darkening skies and mournful sighing winds of autumn herald the approach of winter's cold and storms, to the far-away districts of the southern continent, where it remains until the recurring season brings the desire to return to the place where last year's nest was formed and the young brood was raised. While loving solitude, it is not a morose or monkish species, shunning its kind, but is frequently met with in small companies of five or six individuals, on the banks of some quiet pool in a secluded grove, peacefully gleaning a meal from the yielding soil or surface of the placid water. As they move with a sedate walk about their chosen retreat, each bows gravely to the others, as though expressing a hope that his friend is enjoying most excellent health, or else apologizing for intruding upon so charming a retreat and such select company. At times they run rapidly along the margin of the pond, often with the wings raised high above the back, occasionally rising in the air to pursue some flying insect, which is caught with much skill and agility. The actions are light, quick at times, and graceful, and the bird flies rapidly, its neat plumage showing to great advantage when the wings are outspread, as it skims swiftly over the surface of the water, across open glades, or amid the trunks and branches of the trees. In ad-

39. Solitary Sandpiper.

dition to such places as the one described, the Solitary Sandpiper frequents tidal creeks, and rivulets away from the sea, and occasionally salt marshes; but I have never seen it on the beach, although I believe it does visit the borders of the ocean at rare intervals. It is often seen at high elevations in damp meadows or margins of springs and pools among the mountains, where its low soft whistling note sounds mournfully amid the stillness of the surrounding forest. When startled, as a rule they do not fly far, but settle soon again, and regard the cause of their temporary alarm with a quiet, indifferent gaze. It feeds on insects, larvæ, worms, small crustaceans, etc., such as compose the daily bill of fare of the members of the Snipe family, and when it has satisfied its hunger it will remain standing often up to its breast in the water or drawn into a small compass on the shore. It frequently may be seen walking calmly in the water with slow measured steps, like the heron does when looking for a good place to exercise his piscatory abilities.

Although this species breeds in so many places throughout the United States in the more northern latitudes, its nest and eggs have been rarely met with. Indeed, there really seems to be only one sufficiently authenticated instance of the true eggs having been taken. This was near Lake Bombazine, in Vermont, when Mr. Richardson discovered the nest upon the ground and shot the female as she was leaving it. The eggs were pyriform in shape, light drab in color, spotted with various shades of brown. They measured 1.37 × .95 inches, and resembled those of the Piping Plover (*Æ. meloda*). Mr. Davie also took what he supposed to be an egg of this species in an open field on the borders of the Scioto River, near Columbus, Ohio, but he does not state that the parent was procured or even seen. It is not unlikely that this bird nests in holes in trees. In Alaska this species was seen in small numbers, at Auvik and Nulato, on the lower Yukon, but its nest was not found. In the Argentine Republic the Solitary Sandpiper arrives later than the other Sandpipers, and shunning the plains, frequents similar localities to those it selects in its northern home—small pools sheltered by trees or herbage—and remains in its chosen spot as long as any water is left. It is stated to be more wild and wary in that country, and frequently emits a clear, penetrating trisyllabic cry, especially when on the wing, when it is uttered continually. Mr. Hudson once saw a Solitary Sandpiper and a Blue Bittern living contentedly in company by the margin of a pool in a wood, sheltered by trees and aquatic plants. When not fishing the Bittern would doze on a branch just over the water, while its little companion busied itself upon the margin of the pool. When disturbed they rose together with a strident chorus composed of the harsh notes of the Bittern and the clear, pure whistle of the Sandpiper. These hermits, self-banished from the world and the society of their kind, lived peacefully and contentedly together in a curiously chosen friendship.

### *TOTANUS SOLITARIUS.*

*Habitat.*—Eastern North America generally to the Plains; northward to the Alaskan Peninsula on the west, and the Mackenzie River on the east, and in winter goes south to Bermuda, West Indies, the Argentine Republic, and Peru. Breeding in the northern States and northward of the borders of the Union to limits of forest growth.

*Adult in Summer.*—Crown and nape, streaked with blackish brown and white; back and scapulars, dark olive brown, speckled with white; wing-coverts, uniform dark brown, larger coverts edged with white; primaries, blackish brown; rump and middle tail-coverts, blackish brown, feathers spotted on edge with white; lateral tail-coverts, white, barred with black; middle tail-feathers, bright brown, remainder white, barred with black; throat, pure white; sides of head, neck, and breast, white, streaked with black, and washed with buff on sides of breast; rest of under parts, pure white, barred with black, sometimes indistinctly on the flanks and under tail-coverts; bill, greenish brown; legs and feet, olive green. Length, 8 inches; wing, 5¼; culmen, 1¼; tarsus, 1¼; middle toe, 1.

*Adult in Winter.*—Similar to the summer plumage, but with few white speckles on the upper parts, and neck and breast indistinctly streaked and washed with ash color.

## WESTERN SOLITARY SANDPIPER.

MERELY a race of the Solitary Sandpiper, this bird is not always easily distinguished from the eastern species. Specimens from Arizona seem to be intermediate.

### *TOTANUS SOLITARIUS CINNAMOMEUS.*

*Habitat.*—Pacific Coast eastward to the Plains; northward in breeding seasons to Arctic Circle. "Similar to *T. solitarius*, but larger; wing grayer, the light spots on the back, scapulars, and wing-coverts brownish cinnamon, instead of white or buffy whitish; the sides of the head more whitish, especially on the lores; no well-defined loral stripe."—*Brewster.*

40. Green Sandpiper.

## GREEN SANDPIPER.

NO record is obtainable that this bird has ever been seen alive in North America. It is an Old World species, and is included in our fauna on the strength of a dealer in England having received a skin among a number of American birds from Halifax, Nova Scotia, which fact seemed to prove that the specimen was killed in that province. This is but negative evidence, and hardly of that satisfactory kind as to warrant the adoption of the species into the American fauna. It represents in the Old World our Solitary Sandpiper, and in its general habits the two species closely resemble one another.

The present one, however, deposits its eggs in old nests in trees, not being very particular whose late abode it selects, whether Thrush, Jay, Pigeon, Blackbird, or even that of a Squirrel! Perhaps our own Sandpiper does the same, which may account for the reason that so far authenticated eggs are so rare and the nest so difficult to find. In a *hole* of a broken-down tree (*Populus tremula*), in which the previous year a Flycatcher (*Muscicapa luctuosa*) had built a nest, four newly hatched young were seen to scramble out and hide in the grass, as related by Mr. Hintz, in the *Journal für Ornikologie*, 1862. The eggs are a delicate sea green, spotted with pale purplish gray and dark brown.

The Green Sandpiper is found generally throughout the Old World from the Arctic regions to the

Cape of Good Hope and from the British islands to China.

### *TOTANUS OCHROPUS.*

*Habitat.*—Northern portions of Old World. Accidental in Nova Scotia (?).

*Adult in Summer.*—Head and neck, striped with blackish brown and white, the brown stripes broadest on the crown; upper surface and wings, dark olive brown, spotted with white; upper tail-coverts, pure white, middle feathers barred with black, remainder with subterminal black bar, and broken bar or spots on apical half of outer web; chin, white; rest of under surface, white, streaked on fore-neck and breast with blackish brown; bill, grayish black; legs and feet, bluish gray, green at the joints. Length, 10 inches; wing, 5⅓; culmen, 1⅓; tarsus, 1⅓.

*Adult in Winter.*—Resembles the summer dress, but has the head and neck grayish brown without spots, and the spots on upper parts obsolete or faint; white stripe from bill over the eye.

41. The Willet.

## WILLET.

IT is difficult to define the exact distribution of this species, if we recognize the subspecies, as both are found together in various portions of eastern North America, and in winter plumage they are not distinguishable from each other; but as a general thing, the present bird may be regarded as an inhabitant of the region east of the Mississippi, although it is often found in Texas, and many specimens from Minnesota cannot be separated from it. It is next to the Curlews and Godwits, the largest of the Snipe family, and exceeds in size some of the members belonging to those two genera. The Willet is exceedingly noisy, its shrill cry of *pill-will-willet* being heard (when the birds are present in numbers on the marshes) at all hours of the day and also during the entire night. It is shy and restless, very wary, keeping at a most respectful distance from every object it distrusts, and when in exposed places, such as sand-bars or bare points running out into rivers or ponds, it allows nothing to approach. It usually associates with individuals of its own species, but may occasionally be seen with Yellow-legs, Sanderlings, and sometimes with Gulls. It will not decoy very readily, and when it does come to the stools, it keeps a sharp eye about it, and should it observe anything to excite its suspicion, is off at once, and no amount of whistling can induce it to return. During the breeding season, however, it seems to change its nature and becomes more trustful or less suspicious or more daring, it is difficult to determine exactly which. Should one enter the territory which has been selected for the nest, the Willets set up the most discordant cries, each bird joining in the chorus and hastening to the assistance of the one who seems to consider himself as the most aggrieved, probably because its nest is nearest the object of their dislike. The birds fly around in circles, screaming in their very best style, frequently swooping down at the intruder, and apparently unmindful of all danger to themselves. Even if shot at, though the majority may move off for a little distance, they are apt to return with renewed scoldings and endeavor to drive their unwelcome guest from the vicinity of their nests. The semipalmated foot enables the bird to propel itself through or over the water with facility, and consequently it is a very fair swimmer; and it often may be seen wading up to its breast. When the tide is high and the bars and marshes are covered, and their food unobtainable and hidden from sight, the Willets gather together and stand upon some open ground, resting from their labors, and apparently reflecting on the uncertainties and trials of bird-life. But even if some may have so yielded to "tired nature's sweet restorer," as to pass the bill beneath the feathers on the back and pay no further attention to surrounding objects, let not any one think that it is a favorable time to approach the little company, for there is always "A chield's amang you, taking notes," and as the observer draws nearer than is deemed convenient, a shrill whistle suddenly sounds upon the air, the birds are at once upon the wing, and with loud cries and swift flappings depart for some more secure retreat. On the Atlantic Coast, although still fairly common, this species, like various others of its tribe, is not so numerous as in former years in many localities. It breeds, however, throughout the countries it frequents, in all suitable places, the nest placed in a tussock of grass near water. It is not more than a depression amid the material in which it is found. The eggs, four in number, greenish white or brownish olive, spotted with brown and purple, measure 1.98–2.12 inches in length by 1.46–1.58 in breadth. Crows often pilfer the nest and destroy many eggs. It is said that during the breeding season this species often alights on trees. I have never seen them do this.

## *SYMPHEMIA SEMIPALMATA.*

*Habitat.*—Eastern North America, between the Mississippi and the Atlantic Coast. Breeds, from latitude 56 to Texas. In winter south to the West Indies and Brazil. Accidental in Europe.

*Adult in Summer.*—Head, neck, back, and scapulars, pale grayish brown, streaked on head and neck and spotted and blotched on back and scapulars with black; primaries, black, basal half, white; coverts, grayish brown, greater tipped with white, forming a bar across the wings; rump, grayish brown, some of the feathers edged with whitish; upper tail-coverts, white, barred more or less distinctly on the apical third with brownish black; middle tail-feathers, pale brown, barred with brownish; rest of feathers, pale brown, growing lighter to the outermost feather, which is nearly white; under parts, white, striped with blackish brown on sides and front of neck, and barred on breast, flanks, and under tail-coverts with dark brown—some specimens have a wash of buff on breast and flanks; bill, black; feet and tarsus, bluish. Length, 16 inches; wing, 8¾; culmen, 2¼; tarsis, 2¼; middle toe, 1½. Specimen from New Jersey coast.

*Adult in Winter*—Above, ashy gray; beneath, purest white, washed on front of neck, and sometimes on breast with gray; wings and tail, like summer dress.

## WESTERN WILLET.

THE Western Willet is somewhat larger and grayer, though not always, than the eastern bird. In winter the two forms cannot be distinguished from each other, save possibly by the longer bill of the present one, though this is not always reliable. In the summer dress specimens of this form are found as dark on the back and blotched with black as any of the common species, and as both forms are frequently found together, the attempt to separate them gives very unsatisfactory results. The difference between the birds, such as it is, seems to have been first noticed by Cuvier, who called this one *Speculifera*, but gave no description, and afterwards Pucheran published a description of Cuvier's type, the example being in the gallery of the Paris Museum. This bird is very common in the western part of the United States and in Texas. On the west coast it is one of the most numerous of the Bay Birds, and in the vicinity of San Francisco is plentiful throughout the year, and possibly may breed there. In their habits, mode of nesting, etc., there does not seem to be any difference between the two forms.

### *SYMPHEMIA SEMIPALMATA SPECULIFERA.*

*Habitat.*—Western North America to Mississippi Valley. Occasional on Atlantic Coast. Breeding range supposed to be from the source of the " Saskatchewan to California.

Ch. Breeding Plumage. Differing from *S. semipalmata* in being larger, with a longer, slenderer bill; the dark markings above, fewer, finer, and fainter, on a much paler (grayish drab) ground; those beneath, duller, more confused or broken and bordered by pinkish salmon, which often spreads over or suffuses the entire under parts, excepting the abdomen. Middle tail-feathers either quite immaculate or very faintly barred. Average length of wing, 8.11; tail, 3.29; tarsus, 2.66; culmen, from feathers, 2.46."—*Brewster.*

# WANDERING TATTLER.

THE Wandering Tattler certainly deserves its name, for no species of this family, save those that are cosmopolitan, traverse so much sea and shore during the changing seasons as this one. Along the vast extent of the Pacific Coast it goes from the Equator onward to the Aleutian Islands in the far north, and to the interior of Alaska, where probably it breeds along the banks of the mighty Yukon. It visits also many islands in Behring Sea, and then as the great ice-fields close the waters of the Arctic regions, and the snow covers all the land, away this pretty species flies to the coral islands of the southern seas, where, beneath sunny skies and balmy breezes, undisturbed, it can roam the shores and feed at leisure, no one to "molest or make it afraid." They have a preference for rocky shores, upon which the waves have flung the weeds floating on the bosom of the sea, and feed upon the small crustaceans hidden in them. About the middle of May they appear among the islands of the Aleutian chain, and at the Seal Islands it has been seen in June and July, going farther north to breed; and the places they select for the purposes of incubation may be well within the Arctic Circle. This Tattler is usually solitary, or goes in small companies of three or four individuals, is gentle and retiring, moves easily and gracefully over the rocks, ascending higher and higher as one approaches it from the water side. When alarmed it flies but a short distance, alighting upon some projecting rock, and

42. Wandering Tattler.

quietly looks back upon its pursuer. The note is a shrill whistle, which is uttered when the bird is startled and resembles, as given by Nelson, *tū-tū-tū-tū*. On the lower Yukon it has been seen at Auvik, and also at Sitka. On their return journey southward they reach St. Michael's Island the beginning of August and remain until September, when they depart for milder climes. The nest and eggs have not been procured, so we have no information regarding them. On the Commander Islands, where Mr. Stejneger has seen them, he says it carries its body like the Spotted Sandpiper, but seldom flirts its tail like that bird, nor has it the peculiar movement of the head and neck. Its flight is graceful and rapid, and its voice loud and harsh, almost screaming. It comes to the islands in May, and is seen on the stony beaches in pairs or very small troops.

### *HETERACTITIS INCANUS.*

*Habitat.*—Pacific Coast from Lower California to Aleutian Islands and Norton Sound, Alaska. Also in various islands of the Pacific Ocean, as the Sandwich and Eastern Polynesian groups. Breeds probably in Alaska and Aleutian Islands.

*Adult in Summer.*—Head, neck, and entire upper parts, uniform dark plumbeous; wings, like back; primaries, dark brown; shafts, white; tail, dark plumbeous; superciliary stripe, sides of face, white, narrowly streaked with dark plumbeous; throat, white, spotted with same; rest of under parts, white, barred with dark lead color; bill, brownish black; feet and legs, apparently greenish yellow. Length, 8 inches; wing, 6½; culmen, 1½; tarsus, 1¼; middle toe, 1.

*Autumn Plumage.*—Is ashy plumbeous above, the feathers faintly edged with white; under parts, white, the neck, breast, and flanks washed with pale lead color.

## THE RUFF.

ON various occasions the Ruff has been obtained in different parts of eastern North America, the majority of the specimens probably having been secured on Long Island, formerly the ideal resort of the Snipe and water fowl; but they have all, so far as I am aware, been in fall or winter plumage, if males, or else in the somber livery exhibited by the female. No example, I believe, possessing the ruff and in the breeding dress of the male has ever been taken within our borders. In the Old World it is found from the British islands eastward through Russia and southward to the Cape of Good Hope, in Africa. The Ruff is polygamous and in the breeding season the males collect together to fight for the possession of the females; and although their conflicts appear very desperate and carried on with great energy and determination, the combatants receive little damage. Their excitement is intense; they assume the position of game-cocks, with heads lowered and ruffs expanded, and strike and claw with the feet. Several of these duels may be going on at once, and after a short tussle the weaker retires and the victor awaits another foe. It is said that the same piece of ground (generally an open space, sometimes slightly elevated) is chosen every year, and the battles commence soon after sunrise. When the ruff falls out, the male loses all interest in the Reeves, as the females are called, and pays no attention to his

43. The Ruff.

family. The female takes upon herself all the cares of nesting, incubation, and rearing the young. The nest is placed on the ground in some swamp and lined with grass. The eggs, four in number, are greenish gray, spotted with reddish brown and pale grayish brown. In winter the Ruff goes in large flocks, resorts to the banks of mud, generally inland, and feeds on insects, and other minute organisms common to such localities.

In its summer dress the male is a most conspicuous object in the landscape, its usually very bright colors showing strongly amid the somber hues of the wet swampy lands it frequents.

## *PAVONCELLA PUGNAX.*

*Habitat.*—Northern portions of Old World; occasional in eastern North America. Breeding in western portion of Old World from the Arctic Sea to the valley of the Danube, and east to valley of the Amoor.

*Adult Male in Spring.*—Head, neck, and upper parts, including scapulars and greater wing-coverts, bright chestnut, barred with black; *ruff*, elongated feathers of the neck, same colors; fore-neck, breast, and flanks, dark chestnut, barred with black, with white feathers interspersed; middle of abdomen and under tail-coverts, pure white; lesser wing-coverts, ashy brown; primaries, dark brown, with white shafts; rump, blackish, margined with rufous; tail, barred with black and rufous; bill, blackish brown, fleshy at base; legs and feet, yellow. Length, 12½ inches; wing, 7; culmen, 1½; tarsus, 1¾; middle toe, 1¼.

The above is a description of one specimen, and will answer only for that one, as the variation in color among these birds is endless, no two being exactly alike. Some have the ruff black, others black and white, chestnut, white, white marked with brown or buff, etc., and it is quite useless to try and select any specimen as typical. In winter the male has no ruff.

*Female, Shot on Long Island.*—Crown, blackish brown, streaked with rufous; nape and hind-neck, pale brown, indistinctly streaked with darker brown; back and wings, black, feathers margined with bright buffy brown; rump and upper tail-coverts, same color and markings; tail, barred alternately on middle feathers with dark brown and buff, lateral feathers grayish brown, with subterminal black bar and buffish brown tip;

sides of head, neck, and breast, pale brown, with brownish speckles, and on the breast occasional bars of blackish brown; middle of throat and rest of under parts, white; bill, black; feet and legs, greenish in skin; no ruff present.

44. Bartram's Sandpiper.

## BARTRAM'S SANDPIPER.

THIS trim, graceful, and prettily marked species is generally distributed throughout the eastern portion of North America from the Rocky Mountains to the Atlantic Ocean. It is a bird of the plains and uplands, rarely seen near water, into which it seldom, if ever, wades, and in its habits is more of a Plover than a Sandpiper, frequenting grassy fields and prairie-like stretches, hunting with active steps the insects that form its chief means of subsistence. It has been obtained in British Columbia at Colville Bay by Mr. Lord, and also on the Yukon, but is apparently only an occasional visitor to the Pacific Coast. As it is one of the most delicious of birds for the table, the Upland Plover is eagerly sought for by sportsmen and gunners of every rank and style, and is known to all throughout our land by many names. Beside the one at the head of this article, it is also called Field, Highland, Pasture, Plain, Cornfield, Grass, and Gray Plover, also Prairie Pigeon in the West, and Prairie Snipe, Quaily on the Assiniboin, and Papabote in Louisiana. On the Atlantic Coast, while seen at times in considerable numbers, it is never met with in the great flocks observed in the Western States, like Kansas and southwards, more especially in Texas, where it congregates sometimes in thousands. In the Middle States of the seaboard it is a wary bird, except perhaps during the breeding season, standing erect in the grassy plains on which it loves to dwell, and regarding with watchful eye any intruder upon its ground.

It is almost impossible to approach near to them, if the sportsman is on foot, unless the land be hilly, but they seem to regard a wagon as of no consequence, and by going in ever-diminishing circles, the gunner can be driven close to the birds, who may have been watching his vehicle all the time. They do not appear to associate their two-footed enemy and the dreadfully noisy "black stick" he carries, with the four-footed animal dragging his load over the grass. Where, however, it is not much hunted, as in some of the less peopled localities in the West, or during the breeding season, Bartram's Plover is very tame and gentle, sometimes barely moving out of the way for either man or vehicle. It associates often at this time with the Golden Plover, and others of the family found in similar localities, and may be seen scattered in groups or singly over the prairies. It walks well and gracefully, and when standing erect, it watches some suspicious object, with its slender neck stretched to its full extent and topped by the well-shaped head, the bird seems much taller than it really is.

The note of the Upland Plover is a loud, long yet soft whistle, and can be heard for a considerable distance. As one is walking over the grassy plain, there falls upon his ear this distinct cry, coming from some unknown locality. He stops and listens, and again clear and soft the note is borne to him, this time distinguished as from above. He looks up and sees nothing but the interminable blue, spread all around. But soon, as he continues gazing, a tiny speck is visible that floats motionless along, and from time to time, from out the very heavens there descends the soft note of the Plover's voice. Descrying some suitable ground, the bird begins to lower, and on fixed pinions often at an acute angle, it sails downwards and alights sometimes on the ground or occasionally on a fence or stake. It stands erect and motionless, with its wings raised high above the back, exhibiting the beautiful markings to the greatest advantage, and then slowly folds them into their proper place. If on the ground, it then moves forward slowly and deliberately, nodding at every step as if in emphatic approval of its surroundings and its sagacity in selecting so suitable a spot, and pays its attention to such insects as may catch the eye, uttering at times a peculiarly mournful sound, quite different from its usual flute-like cry, to be answered possibly from out the heavens above by some comrade not yet distinguishable to the naked eye. The flight of the Upland Plover is well sustained and swift, and often performed (as will be imagined from the above) at a great height, indeed so lofty at times that its voice alone indicates its presence, the bird being fairly out of sight. It will alight indiscriminately on the ground, fence, telegraph pole, or, as has been noticed, even on a *barn*. When mated the pair keep close company, seeking food together, and are rarely separated by any distance. The nest is placed on the plain or prairie in some open spot, frequently near some water. It is not much of a structure, just a little grass in a depression of the ground; but almost impossible to find at any time, even when the bird is on the eggs, so admirably does her plumage harmonize with that of her surroundings. The eggs, four in number, are clay color, spotted all over with dark brown, and purplish gray. They are very large for the size of the bird, measuring on an average about 1¾ by 1¼ inches. In June the young appear, curious, puffy, overbalanced little creatures, stumbling about the prairie, and are objects of great solicitude to the anxious mother. Should any one approach near her charge, all the artifices known in birdom are practiced to lure the intruder away, while the clumsy little downy balls at the first cry of alarm strive to hide themselves away in the grass. At this time the old bird is perfectly fearless, and will fly close to one, trying her utmost, poor defenceless creature, to drive and frighten away the object of all her fears.

In certain places on the prairie sometimes quite a number of these birds with their young will gather together, and then if one draws near, the scene above described will be repeated on a larger scale, the air being filled with flying scolding birds, swooping in all directions and dinning one's ears with their shrill, loud cries. In the beginning of August this species frequently deserts the prairie and resorts to the plowed land, and is then very wild and difficult of approach. Its food is chiefly insects, and the birds eat a great many grasshoppers. The species will also swallow various berries. In September old and young move southward, sometimes in large flocks. On the pampas of the Argentine Republic, its winter home, it begins to arrive in September and is said to scatter, with a very even distribution, over an area 50,000 square miles in extent. So well are the birds distributed over the pampas, it has been imagined, strange as it may seem, that each one occupies the same range it did the year before, if happily it may have lived to return. It is called there *Chorlo solo* or *Batitú*, and is very shy, hiding in the grass or crouching close to the ground like a snipe. It remains on these vast plains until March; but the return journey to its far northern breeding place is commenced in February, and for two months, high in the air, day and night, is heard the flute-like cry of this Sandpiper as in serried ranks it moves onward to the land where the buds are beginning to burst and all nature is rousing herself from the winter's sleep.

### *BARTRAMIA LONGICAUDA.*

*Habitat.*—Eastern North America. In the north it goes to Nova Scotia and the valley of the Yukon, breeding in its northern range; south in winter to Argentine Republic and Peru. Accidental in Europe and Australia.

*Adult.*—Crown, nape, back, and scapulars, black, the feathers margined with bright buff; hind-neck, bright buff, streaked with dark brown; lesser wing-coverts, pale brown, barred with black and bright buff; greater coverts, dark brown, barred with white on inner web and margined and tipped with same; primaries, dark brown, margined with pale brown or white; lower back, rump, and central tail-coverts, black; lateral coverts, black at base, rest buff, barred with black and white; middle tail-feathers, slatey gray, barred with black; remainder, buff, with subterminal bar and anterior spots of black and tipped with white; sides of head and neck, bright buff, streaked with dark brown; throat, breast, and entire under parts, creamy buff, with irregular dark-brown bars on breast and flanks, some on lower part of breast, arrow-headed in shape; bill, brownish black, yellowish at base; feet and legs, yellowish. Length, 12 inches; wing, 6½; culmen, 1¼; tarsus, 2; middle toe, 1.

## BUFF-BREASTED SANDPIPER.

THE Buff-breasted Sandpiper is more nearly related to the Upland Plover than to any other member of this family. In its habits it resembles the larger species, and like that bird prefers fields and grassy plains rather than the wet and swampy lands frequented by other Sandpipers. It is not common on the eastern coast, but seems to be more an inhabitant of the interior. At the same time, small flocks of about half a dozen are occasionally met with in various parts of the Atlantic seaboard. It is usually very gentle, and pays but little attention to any one who may be near, watching the active creature as it runs about the shore intent only upon securing a meal. It goes far to the north, breeds at Point Barrow in the Alaskan Peninsula, and has been seen on the Yukon at Nulato, and also is very common in the Anderson River region, where it also breeds. According to Murdoch this species arrives at Point Barrow by the middle of June, and the birds spread themselves over the dry parts of the tundra. During the breeding season they indulge in curious movements, one of which is to walk about with one wing stretched out to its fullest extent and held high in the air. Two will spar like fighting-cocks, then tower for about thirty feet with hanging legs. Sometimes one will stretch himself to his full height, spread his wings forward and puff out his throat, at the same time making a clucking noise, while others stand around and admire him. They are silent birds at all times, and the breeding sea-

45. Buff-breasted Sandpiper.

son over, they quietly disappear, never assembling in flocks, and by the beginning of August all have gone south. The nest is, like those of most waders, merely a depression in the ground, lined with a little moss, and the usual complement of four eggs is deposited with the small ends down. They are ashy or olive drab, blotched with various shades of brown, and stone-gray underlying markings, smaller at the pointed, larger and more confluent at the rounded, end. When fired at, this Sandpiper will fly but a short distance, performing a half circle along the shore, and alight again near to the place from which it started, or if on the plain, drop down again at a little distance and run about seeking for insects without exhibiting any signs of alarm. Its note is low and weak, merely a *tweet* once or twice repeated. In winter it goes as far south as the Argentine Republic, which it reaches in October, and lives on the pampas in small companies, associating with the Golden Plover and other migratory species. It may go to Patagonia. In May flocks of two to five hundred proceed north, the birds flying low and continuing to pass at intervals in certain localities for several days.

### *TRYNGITES SUBRUFICOLLIS.*

*Habitat.*—North America, breeding in Alaska and British America. South in winter to Uruguay and Peru; occasional in Europe.

*Adult.*—Entire upper parts, pale clay buff, every feather with the center black; inner secondaries, lustrous blackish brown, margined with brownish buff; primaries, dark brown, blackish towards the tip and edged with white, and white on the outer edge of the inner webs, mottled with black; larger under wing-coverts, marbled with black and white and subterminal black bar and white tips; middle tail-feathers, dark brown, with a greenish tinge, remainder pale brown, with a subterminal black bar and tipped with buffy white; entire under parts, pale buff, almost white on abdomen; feathers with white tips, and those on sides of the breast with black central spots; bill, greenish black; legs and feet, yellowish green. Length, 7½ inches; wing, 5¼; culmen, ¾; tarsus, 1¼; middle toe, ⅞.

The marbling on inner web of the primaries varies considerably in size, those specimens from California having it apparently much coarser, but I have not a sufficient series at hand to determine whether or not it is constant.

46. Spotted Sandpiper.

## SPOTTED SANDPIPER.

ONE of our best known and familiar birds, the Spotted Sandpiper, is met with on the shores and banks of nearly all our lakes and rivers, which it enlivens with its sprightly presence, and draws attention to itself by its soft note and the curious balancing of the body as it stands upon a stone near the water, or even when walking sedately along the topmost rail of a fence. It is a very common species, and is distributed generally throughout North America from the Atlantic to the Pacific Oceans, and breeds wherever it is met with in the springtime. It arrives in April from the far southern lands where it has passed the winter, and soon commences the courtship preparatory to the nesting season. The "Tilt-up," or "Peet-weet," as it is also called, does not go in flocks of any size, but is rather solitary in its disposition, an individual or pair seeming to appropriate a certain amount of the shore, where they dwell contentedly, only flying, when disturbed, higher up or lower down the river, as the case may be, and then if any Tiltup is on the particular spot near where they desire to alight, they move on to some other part of the bank or beach. The flight is rapid, performed with quick, stiff beats of the wings, and the bird utters frequently its cry of *pēēt-wēēt* as it passes along. It is a most comical species to watch upon the shore. When it alights, after its short flight, it may stand for an instant motionless contemplating its surroundings, and then makes a profound bow, inclining both head and neck, at the same time elevating its hindquarters in a seeming derogatory manner, very disrespectful to the onlookers, and as if to emphasize the fact that the motion was intended for each and all of those present, it deliberately moves around on its feet, presenting head and tail alternately to first one point of the compass and then to another. It is usually silent during this performance, its importance and solemnity doubtless precluding any such thing as idle remarks. So long as it remains upon the shore these depressions and elevations of alternate ends occur frequently, and sometimes the bird stops even when in chase of some elusive insect to repeat this mark of its distinguished consideration for its observer. The nest, lightly built of straws and grasses, is placed in open spots, either along the borders of streams or ponds, or in fields among the stubble. The eggs are light drab, or cream color, spotted with purplish brown or other shades of the latter color, the spots becoming blotches and confluent at the larger end. The young run as soon as hatched, and are great adepts at hiding on the approach of danger. When alarmed the mother shows great excitement, and the frequency and violence of the movement behind and before, already remarked upon, is extraordinary, as she runs about with plaintive cries of *pee-weet-weet*, and with outspread wings strives to draw attention to herself. The "Teeter Snipe" rarely flies any distance in a straight line, generally on a curve, sometimes in a zigzag course, and during the nesting season the male utters what may be really called a song. As he flies over the surface of the water, he will suddenly check his course and rise upwards for a short space on quivering wings, pouring out rapidly repeated *peet-weets* in an earnest, trilling stave of snipe music, as if his joy in life and appreciation of his present surroundings and prospects were altogether too much for his feelings, and greater than his small body could possibly contain. Then, having relieved his surcharged breast by these liquid notes, he continues his course to the nearest beach or stone and gravely bows his approval of the whole matter. When passing each other on the wing the Peetweets almost always exchange salutations and personal remarks in their cheerful, whistling notes, keeping up the conversation often after considerable distance has intervened between the travelers. This species is as much at home upon a fence-rail, hay-stack, or stake as on the ground, and frequently alights on one of these to survey its surroundings. The female with young seems to select such a place in order more readily to perceive approaching danger from a distance, and also probably as more convenient to keep an eye upon her lively, active little downy offspring, as they scramble over the ground chasing the various insects that attract their eyes. In the Rocky Mountains this species is found at high elevations, even up to the limit of timber, and is as much at home amid such lofty heights as at the level of the sea. It is an interesting, harmless creature, not very suitable for the table (although many are shot for food), and in its demure, attractive dress of quiet colors, pretty ways, and soft voice, it is one of the most pleasing objects seen along the borders of our rivers in the summer time.

*ACTITIS MACULARIA.*

*Habitat.*—North and South America to Brazil; less common on Pacific Coast. Breeding throughout temperate North America. Occasional in Europe.

*Adult.*—Entire upper parts, lustrous ashy green, spotted irregularly with brownish black; the neck with less greenish lustre and more of an ashen hue; white stripe over the eye, reaching to the nape; entire under parts, pure white, with numerous brownish black markings over all the surface, the spots smallest on the throat; primaries, dark brown, white on basal portion of inner web; tail, ashy green, remainder with subterminal blackish bar, outer with dusky and white transverse spots, and all except central pair tipped with white; bill has edge of maxilla and mandible yellow, remaining portion black; legs and feet, grayish olive. Length, 7¾ inches; wing, 4; culmen, 1; tarsus, 1; middle toe, ¾.

47. Long-billed Curlew.

## LONG-BILLED CURLEW.

FROM British North America to the Gulf, and from the Atlantic to the Pacific, the Long-billed Curlew is met with generally, although in not such large numbers in the Eastern States, where in many localities its appearance is by no means regular. It is more abundant in the South, and I have met with large flocks in Eastern Florida. It is the largest of the waders, and is always a conspicuous feature of the locality it frequents, presenting a rather curious appearance with its long curved bill of such unusual proportions. Unlike the other species of the same genus, the "Sickle-bill" does not migrate far to the north, but remains in the temperate regions of the continent, breeding in quite low latitudes. It is found throughout the Mississippi valley, and is common on many of the vast grassy plains of the West, frequenting both dry and wet localities, breeding in nearly all its places of resort. The flight is strong and well sustained, the members of the flock proceeding, after the manner of geese, in a triangular order, some old bird at the apex, heading the procession, and uttering at intervals a hoarse cry, understood and obeyed by those following. They readily respond to an imitation of their call, and wheel to the decoys, approaching on widespread motionless wings, each bird presenting so large and steady a mark that there is no excuse for the sportsman to miss his aim. No species among the waders is more sympathetic, or evinces greater solicitude for its companions if any are in distress, than this Curlew. Once when shooting in Florida, in the vicinity of St. Augustine, a large flock of these birds passed overhead, and I brought down some by two shots from my gun. Although naturally much alarmed, the survivors immediately returned to their wounded companions, which were calling aloud as they lay upon the marsh, flying over and around them, with hanging legs, and uttering answering notes of sympathy, and approaching nearer and nearer until they were not many feet above the ground. Repeated discharges of my gun failed for a time to drive the unwounded birds from the vicinity, but as each individual fell from their ranks, the rest would swoop towards it, and with much crying seem to urge it to rise and follow them. The air was full of rapid-flying circling birds, each one screaming its best, and it was not until a considerable number had fallen that the remainder, convinced at length of the fruitlessness of their efforts, and the danger present to themselves, departed for a more secure locality. Some of the specimens obtained at this time were very large, with rather uncommonly long bills. The flesh was rank and the bodies gave out an "unpleasant fish-like smell."

Although coming so readily to decoys, this Curlew is a shy and wary bird, very watchful when standing on the open plains, and permits no one to approach near it, easily taking alarm and flying off with loud cries. It has many names in different parts of the United States, such as Hen or Buzzard Curlew, Smoker, Sabrebill, and in the south, Spanish Curlew, and in parts of the New Jersey coast, Mowyer. While the Sickle-bill is a dweller of the marshes, it by no means confines itself to them, but often visits dry sections of the country. The food consists of various small shellfish, worms, insects of different kinds, and berries. The flesh is more palatable when the birds frequent the interior, especially when its supply of food is mainly that of berries. The nest is merely a depression in the ground lined with a few grasses, and the eggs, usually four in number, are shaped something like that of a common fowl, clay-colored, with olive or buff shades, spotted with sepia or chocolate, the markings small and regularly distributed. They resemble the egg of the Willet somewhat. During the breeding season it is very solicitous of its eggs, or young, and its loud, harsh cry, when any one invades the territory where its family affairs are progressing, can be heard for a long distance. In Manitoba this species is rare, and on the Pacific Coast it has not been found north of Vancouver Island. Its southern range is Central America and some of the West India islands, but I know no record of its occurrence in South America. On Long Island it appears after the breeding season, apparently migrating northward for a brief period.

### *NUMENIUS LONGIROSTRIS.*

*Habitat.*—North America from the Atlantic to the Pacific in the temperate regions, going south in winter to Cuba, Jamaica, and Central America. Breeds in the interior of its northern range and in the South Atlantic States.

*Adult.*—Entire upper parts, pale rufous, most reddish on the back and scapulars, each feather with transverse confluent black bands, inclined to streaks on crown and neck; primaries and secondaries, cinnamon, with blackish bars; outer webs of primaries, blackish brown; wing-coverts, paler than back, blackish brown in the center, variegated on rest of feathers with rufous and ashy white; upper tail-coverts, rufous, barred with blackish brown, whitish near the ends of the feathers; tail, rufous, washed with ash and crossed with blackish-brown bars; throat, buffy white; rest of under parts, buff; rufous on the flanks; the neck streaked, and flanks barred with brownish black; bill, black, fleshy brown on basal half of mandible; legs and feet, grayish brown. Length, about 2 feet; wing, 10 inches; culmen, 8, varying greatly; tarsus on top, 2¼.

Individuals vary greatly in the depth of the rufous coloring of their plumage, some being quite pale and others dark cinnamon, and there is also great difference in the length of the bills, some being moderate in this respect, while occasionally an individual is met with having one enormously lengthened. Both color of plumage and length of bill are evidently purely individual variations.

48. Hudsonian Curlew.

## HUDSONIAN CURLEW.

UNLIKE its larger relative, the Long-billed Curlew, the present species migrates in the different seasons to the Arctic Sea on the north, and to the plains of Patagonia on the south, traversing the entire length and breadth of the continents of North and South America. In some localites it is at times quite numerous, but I have never seen it as abundant as the Sicklebill, and have regarded it, at least on our eastern seaboard, as not a very common bird. It frequents the marshes and muddy flats, associating with the Willet and Godwit, and feeds on worms and various shellfish.

The Jack Curlew, as it is generally known, breeds in the far north in the lands bordering the Arctic Sea, on the Barren Grounds in the Anderson River region, and in the northwest, on the open country about the shores of the Polar Sea. In the island of St. Michael's they are common about the beginning of June, but always shy. They do not remain there long, but pass on still farther north, returning again, the season of reproduction ended, about the 1st of August, remaining throughout the month, feeding on blue and crow-berries, until they become very fat and heavy. In September it moves southward on its long journey to the confines of South America, migrating along the Alaskan coast by way of Sitka and the Pacific shores of the United States. Many also pass through the valley of the Mississippi, but I consider it the least common of our Curlews in the United States. About the mouth of the Yukon in spring it is not rare, and may often be seen standing on one leg on some stump or log. In the Anderson River region where MacFarlane observed it breeding, the nest was as usual a mere depression in the ground lined with a few withered leaves placed near small lakes or streams. The eggs, four in number, were a creamy drab, spotted with slaty brown, but varying considerably both in ground color and markings, and were larger than those of the Eskimo Curlew (the next species), measuring from 2.21 to 2.40 inches in length by from 1.57 to 1.65 in breadth. The nests from which these eggs were taken were found in about 70° north latitude.

This species bears many names among the gunners in various parts of our country, among which I may cite Short-bill Curlew, Crooked-bill Marlin, Striped-head, American Whimbrel, Horse-foot Marlin, and Jack. On the coasts of New Jersey and West Virginia it used to be very abundant, but of late years, as is the case with the rest of the waders which formerly enlivened our coasts with their graceful forms and attractive dress, these birds appear in constantly decreasing numbers.

## *NUMENIUS HUDSONICUS.*

*Habitat.*—North and South America and West Indies. From the Arctic regions to Patagonia in winter. Breeding range from Greenland to Alaska.

*Adult.*—Crown, blackish brown, with a central stripe of buff; dark-brown stripe from bill through eye to ear-coverts; superciliary stripe, pale buff; neck, pale buff, streaked with blackish brown; upper parts, sooty brown, spotted with whitish buff; wing-coverts, similar but lighter; rump and upper tail-coverts, dark brown, spotted with dark buff, slightly rufous; tail, rufous, barred with dark brown; primaries, blackish brown, shafts white on first two; sides of head, neck, and entire under parts, light buff, marked with narrow streaks on sides of head, neck, and breast, and barred on flanks, with dark brown; bill, brownish black, flesh color at base of mandible; feet and legs, black. Length, about 17 inches; wing, 9; culmen, 3½; tarsus, 2¼; middle toe, 1⅓.

# ESKIMO CURLEW.

IN the Mississippi Valley this species is the most abundant of the Curlews, and in immense numbers scatters over the prairie in every direction, associating with other species of its tribe which frequent similar localities, such as Bartram's Sandpiper, Golden Plover, etc. When feeding about in such large flocks, they keep up a constant low chattering noise, as if indulging in an uninterrupted flow of conversation. On the Atlantic Coast it does not appear in such great flocks, so far as I am aware, and although in certain portions of the eastern shore it is more numerous than the Hudsonian Curlew, it is not usually very abundant. It frequents the open flats in the vicinity of the seashore, feeds on insects, worms, etc., and is a shy bird, and in the autumn becomes very fat and its flesh is highly esteemed. It is known as Doe Bird, Futes, Small Curlew, etc. It flies with great rapidity, is easily alarmed and difficult to approach, unless when feeding quietly among other waders, but is more readily shot when flying to and from its feeding grounds, when if one stations himself on their route, as they generally pass at no great height, many can be secured. It passes north in May and returns along the eastern coast in August, not making a lengthy stay in any locality, but going rapidly on to its winter home in the far southern parts of South America. It generally arrives from the north in August after an easterly storm, but is irregular in its movements, and in some seasons is much more abundant

49. Eskimo Curlew.

than in others. In Labrador this Curlew is seen in flocks of various sizes ; sometimes several thousand of the birds are gathered together, and their flight is swiftly performed by regular beat of the wings, and they often execute many beautiful evolutions, frequently massing together in compact ranks. On alighting the wings are raised over the back, as is the habit of many Snipes and Plovers, and then folded carefully and with deliberation into the accustomed place. The note is a soft, clear whistle, and the birds come readily to the gunner (if he can imitate their call cleverly), dropping the legs and curving the wings as they sail unsuspectingly to the decoys. They feed on grasshoppers, berries of various kinds, and small snails which they detach from the rocks. In Northern Alaska the Eskimo Curlew is abundant along the coasts of Behring Sea and Kotzebue Sound. It has been obtained on the Yukon and at Point Barrow. They go in flocks of from twenty to one hundred and fifty, and pass the island of St. Michael's in May to the breeding grounds within the Arctic Circle. Nelson says that small flocks of this Curlew will follow a single Hudsonian Curlew all over the country in the same manner as smaller species of Snipe will follow one of a larger kind, and he imagines it is on account of their dependence on the superior watchfulness of the larger bird, and a greater degree of protection thereby secured. On the Barren Grounds up to the Arctic Sea this Curlew breeds, and MacFarlane found the nest in the Anderson River region. It was merely the usual depression in the ground, lined with a few decayed leaves and dried grass. The eggs vary in dimensions and coloration, being either green, gray, or brown, marked with different shades of sepia in various size spots, and measure on an average from

2.04 by 1.43 inches. This Curlew, like its larger relatives, is very sympathetic in its disposition, and lingers around the place where its companions have fallen victims to the sportsman's wiles, often paying with its life for the utter disregard shown for its own preservation. In the Argentine Republic it arrives on its long journey from the extreme north in September and lingers until late in February, dwelling on the pampas in company with its friends, the Golden and Upland Plovers.

### *NUMENIUS BOREALIS.*

*Habitat.*—North America, migrating in winter to extreme point of South America. Breeding in the Arctic regions from Greenland to Behring Straits. Accidental in Great Britain.

*Adult.*—Closely resembles the Hudsonian Curlew, but is darker upon the back, and is much smaller in size; top of head, black, streaked with buff; black line from bill through eye to ear-coverts; rest of head and entire neck, buff, streaked with blackish brown; upper parts of body and wings, blackish brown, feathers margined with buff, lighter on wing-coverts; primaries, dark brown; upper tail-coverts, like the back, but showing more buff; tail, grayish brown, barred with dark brown; throat, buffy white; breast and rest of under parts, buff, streaked on breast and barred on flanks with blackish brown; bill, brownish black; base of mandible, flesh color; legs, greenish brown. Length, 13½ inches; wing, 8; culmen, 3; tarsus, 2.

50. Bristly-thigh Curlew.

## BRISTLY-THIGHED CURLEW.

A NATIVE of various islands in the Pacific Ocean, only two specimens of this Curlew have been taken on our Western Coast, one on St. Michael's Island and the other on Kadiak Island, Alaska. The first-mentioned individual, which was the second one procured, was killed by Nelson as it stood near where he was shooting Black Brant. There were two together, and they uttered a loud whistle similar to the Hudsonian Curlew. He shot both, but one, unfortunately, was lost in the grass where it fell. Its appearance on any part of our shores can only be regarded as purely accidental, its native islands lying several thousand miles from our western coast. The species is peculiar in having the shaft of the thigh feathers extended beyond the webs and resembling bristles. The examples enumerated are all that have been obtained within our limits, so far as I am aware.

### *NUMENIUS TAHITIENSIS.*

*Habitat.*—Islands of the Pacific. Occasional in Alaska.

*Adult.*—Top of head, sooty brown, with central stripe of buff; a blackish streak from bill through eye; rest of head and neck, buff, streaked with dark brown; superciliaries, buffy white, streaked with brown and reaching to nape; back and scapulars, chocolate brown, with large spots of buff; wing-coverts, like scapulars, but paler; rump, similar to back; upper tail-coverts and tail, reddish buff, the latter barred with dark brown, the coverts sometimes marked with brown; throat and under parts, buff; the neck and breast streaked, and flanks barred, with dark brown; shafts of tibial and femoral feathers lengthened like bristles; bill, horn black; base of mandible, dull flesh color; feet and legs, livid blue; iris, hazel. Length, about 17 inches; wing, 10½; tarsus, 2⅓; culmen, 3⅝.

## THE WHIMBREL.

ESSENTIALLY an Old World species, this Curlew is only admitted into our fauna from the fact that it occurs occasionally in Greenland. In its habits and nesting it bears a close resemblance to other species of Curlew.

### *NUMENIUS PHÆOPUS.*

*Habitat.*—Northern parts of Old World. Occasional in Greenland.

*Adult.*—Top of head, sooty brown, with a central stripe of whitish buff; dark-brown stripe from bill through the eye; superciliary stripe, extending to the nape, buffy white, streaked with brown; neck, whitish, streaked with dark brown; back and scapulars dark brown, feathers margined with grayish; primaries, blackish brown; rump, white; upper tail-coverts, white, barred with dark brown; tail, grayish brown, central pair darkest, barred with dark brown and tipped with white; throat and entire under parts, white, neck and breast streaked with dark brown, flanks and under tail-coverts barred with the same; bill, black; base of mandible, pale brown; legs and feet, grayish blue. Length, 17 inches; culmen, 3; wing, 9¼; tarsus, 2⅓.

51. The Whimbrel.

52. The Lapwing.

## THE LAPWING.

NAMED from its flapping mode of flying — also sometimes called, from its cry, the *Peewit* — the Lapwing is one of the most common and familiar species of the Old World in its northern regions. It ranges from the British islands to Japan, and in winter goes south into Northern Africa. During the breeding season the male has a curious habit of flying near the nest and throwing himself about in the air, as if his senses had left him, uttering all the time a peculiar wailing cry. The eggs, always four, are brownish olive, spotted and blotched with purplish and blackish brown. The nest is the usual hollow in the ground, sometimes bare, or lined with leaves and small sticks.

The Lapwing is an occasional visitant to Greenland, and has been taken on Long Island near New York. Dall mentions the capture of what he supposed to be this species (he did not see the specimen) on one of the small islands in Norton Sound, Alaska, off the mouth of Golsova River. These are its only claims to admission in our fauna.

### *VANELLUS VANELLUS.*

*Habitat.*—Northern parts of Old World. Occasional in Greenland, Long Island, New York, and on the islands in Norton Sound, Alaska.

*Adult Male in Spring.*—Forehead, lores, top of head, throat, and breast, velvety black, with a purplish lustre; feathers of occiput elongated into a crest, curving upwards, also velvety black; nape, side of face, and side of neck, white, marked with black streaks behind the eyes; back, scapulars, and inner secondaries, metallic green, blotched with bright purple; wing-coverts, metallic purplish violet; primaries, purplish black, the three first terminating in white; rump, metallic green, with blue reflections; upper tail-coverts, chestnut red; tail, white on basal half, rest black, this decreasing towards the outer feather, which is sometimes altogether white, and the tips of black feathers white; under parts below the breast, white; under tail-coverts, rufous; bill, black; legs and feet, dull lake red. Length, 13 inches; culmen, 1; wing, 9; tarsus, 2.

*Female.*—Like the male, but with the throat white, crest shorter, and the upper parts generally duller in color.

*Male in Winter.*—Differs from the summer dress, in having a white throat, the black pectoral band tipped on some feathers with white, and also some on the back tipped with buff.

53 Black-bellied Plover.

## BLACK-BELLIED PLOVER.

IN some of its stages of plumage the Black-bellied Plover, or Black-breast, is frequently confounded with the next species, or European Golden Plover, but can easily be distinguished at all ages by having the axillary plumes (the long feathers growing from the arm-pit and seen underneath the wing,) *black*, whereas in the other species these are *white*. This species passes through the United States during migration, going northward to its breeding ground in the month of May and returning in August. They are more numerous along the seacoasts than in the interior, although at times in Manitoba, and some of the States in the Mississippi Valley, they appear in large numbers, as mentioned by Mr. Cooke, when he once observed them congregated in thousands on the Platte River in Nebraska. The young are so different from the adults in plumage that many look upon them as belonging to another species, and they are called Bull-headed Plover, or Beetle-headed Plover. On the Pacific Coast, in the Northwest, this bird is a rare visitor at Point Barrow, but is rather common on the Yukon, and also appears at Sitka during the migration. About the end of August they reach the mouth of the Columbia River, and later on the California Coast, and are abundant in the northern part, more so than in the southern end of that State. The Gray Plover, (the name by which its relative is usually known in Europe, where it is also called, especially in the British islands, Sea-cock, and Sea,

Strand, Mud, Stone, and Rock Plover,) is a very handsome bird when arrayed in its summer plumage, its showy white back, dotted with black, affording a strong contrast to the jet black of the under surface. It is not, however, so well known to the majority of persons in that dress, the winter robe, or one similar to it borne by the young, being most familiar. Along the coast this Plover frequents the salt marshes and flats left bare by the tide, as well as the shores of ponds, feeding upon insects and small shellfish, and in such places its flesh begets a fishy or sedgy flavor; but on the uplands, where it feeds on berries, grasshoppers, and such like objects, it is more palatable. In the spring it is rather a shy bird and does not come readily to the decoys, but in the autumn on its return, when it goes in larger flocks, it is tame, especially if there are many young birds associated with the old ones, and then decoys readily and affords excellent sport. It is a beautiful sight to witness a flock come to the decoys, sailing steadily along on motionless decurved wings, the birds lisp their gentle note of welcome or inquiry to the wooden representatives scattered about the meadow, until, hovering above them for a moment with hanging legs preparatory to alighting among their supposed fellows, the deadly discharge rings out from the gun of the concealed sportsman, and the ground is strewn with hapless victims, while the survivors, startled, spring upwards, and with rapid wing-beats hastily leave the dangerous spot. The Black-breasted Plover breeds in high northern latitudes, and its nest and eggs have been found by several enterprising travelers in those unattractive regions. The nest is but a depression lined with grass or moss, and the eggs, usually four in number, vary greatly in color, from brown to a greenish drab, spotted and blotched with rufous brown distributed generally over the entire shell. They are pyriform in shape, and vary from 2–2.30 inches in length and 1.40–1.47 in breadth. Its breeding range in America seems to be Melville Peninsula and the Anderson River region, and in Alaska about the Yukon; probably also in other localities between these extremes. Besides the names already given it is called May Cock in Massachusetts and Plot in Virginia, at Cobb's Island.

### *CHARADRIUS SQUATAROLA.*

*Habitat.*—Northern part of northern hemisphere, migrating in America southward in winter to the West Indies and Brazil. Breeding in Northern Siberia, Alaska, Anderson River region, and Melville Peninsula.

*Adult Male in Summer.*—Lores, sides of face, neck in front from hind part of eye, and under part of body, jet black; front and top of head above eyes, nape, back of neck, widening on the sides to breast, white; centre of crown with some black feathers showing; back and scapulars, black, spotted and barred with white; wing-coverts, ashy brown, spotted and barred with white; greater coverts, ashy brown, margined with white; primaries, blackish brown on outer webs, white in center of inner webs, and with white shafts; under wing-coverts white, axillary plumes black; rump, brown, feathers margined with white; upper tail-coverts, pure white, barred irregularly with brownish black; tail, white, barred with brownish black; vent and under tail-coverts, white; bill, black; legs and feet, grayish white. Length, about 11 inches; wing, 7½; culmen, 1¼; tarsus, 2.

*Adult Male in Winter.*—Upper parts, dark brown, with irregular white markings, most numerous on the wing-coverts; upper tail-coverts, white; entire under parts, white; lower part of neck and breast, mottled with pale brown, and the under tail-coverts slightly marked with brownish black; axillary plumes, black.

## EUROPEAN GOLDEN PLOVER.

THE European Golden Plover resembles the American species so closely that it might easily be mistaken for it by any one not an expert. The only difference appearing to be reliable between them is, that while the European has the under wing-coverts and axillary plumes white in the adult, or nearly so, the American has these smoky gray. In their habits there is no appreciable difference. This bird is said to occur in Greenland and even to breed there, hence its claim to admission into our fauna. Its chief breeding places are in the northern parts of Great Britain, Norway, Russia, and Siberia east to the Yenisei, migrating in winter sometimes as far south as the Cape of Good Hope. In the Yenisei valley an allied species, the Pacific Golden Plover, breeds.

### *CHARADRIUS APRICARIUS.*

*Habitat.*—Europe, in winter to Africa; Eastern Greenland. Breeding in Northern parts of Old World from the Atlantic Coast to the valley of the Yenisei in Siberia.

*Adult Male in Summer.*—Top of head and entire upper parts, black, spotted all over with bright yellow, or golden, and white; wing-coverts, dusky brown, spotted sparingly with yellow and white, the latter most conspicuous; primaries, rufous brown, with white shafts; upper tail-coverts, black, irregularly barred with gold: tail, dark grayish brown, barred with white, and a few tinges of yellow; line across forehead, passing over the eyes and extending down the sides of neck, widening as it goes, until it forms a broad patch on side of breast, and includes the flanks, white, bordered on breast and flanks with black, spotted with gold; sides of head and neck, throat, jugulum, middle of breast, and under parts, jet black; under

54. European Golden Plover.

tail-coverts, mostly white; axillary plumes, white; bill, black; legs and feet, bluish gray. Length, 10 inches; culmen, 1; wing, 7; tarsus, 1½.

*Adult in Winter.*—Upper parts as in summer, the gold spots varying in amount among individuals; under parts, white; lower part of neck and chest mottled with grayish brown and gold, the latter most conspicuous on the sides; under wing-coverts, white, slightly mottled with brown.

## AMERICAN GOLDEN PLOVER.

IN the United States the Golden Plover or Green-back is only a migrant, passing in greater or less numbers in spring and autumn on the way to and from its remote northern breeding grounds. It goes mostly by the seacoast, or if the weather is favorable, far out at sea, making but few stops in the long journey. In the autumn I have met with it in great flocks on the prairies of Illinois, where they were scattered over the grassy plains, feeding on insects and such other food as abounded, running lightly about and uttering their mellow note. At times they would rise simultaneously, as if by some preconcerted signal, mass together in one great body and speed over the ground with wonderful rapidity, executing intricate evolutions, wheeling and darting hither and thither, as if the entire flock was actuated by a single impulse. At such times when the flock passed low over the earth, presenting the flank of the compact mass, numbers would fall at the discharge of my gun, but the report did not seem to especially alarm the rest, for after proceeding a short distance the birds would wheel, return over nearly the same route, and pass again as near as before. They would repeat this perhaps several times and then alight on the prairie, raising their wings over the back for a moment before folding them away, stand motionless for an instant as if reconnoitering their surroundings, and then scatter in pursuit of food. They were fat and in excellent condition, and most palatable. In the spring this

55. American Golden Plover.

Plover seems to have no general migratory route, but in the autumn more pass along the seacoast than through the interior, and great distances are accomplished without a stop being made. Frequently going directly out to sea, they fly to the eastward of the Bermudas, if the weather is favorable, and so by way of the West Indies to South America. But should an easterly storm arise they are blown back and then appear on various parts of the Atlantic Coast in vast numbers. This Plover has many names beside those already given, such as Green-head and Green Plover, Pale-breast or Pale-belly, Golden-back, Frost Bird, Squealer, Toad-head, Prairie Pigeon, etc. It often goes in company with different species of waders, especially the Red-breasted Snipe, and is a less timid bird than the Black-breasted Plover. It does not frequent much the wet grounds, preferring rather high and grassy plains, and on the seacoast is always found back from the shore in the fields and open level stretches. It runs rapidly and frequently for a considerable distance when alarmed before taking wing. When first disturbed all the individuals in a flock stand silent and motionless, watching the object of their suspicion. At such times it is rather difficult to see them, their plumage harmonizes so well with the ground. Then, when that which has alarmed them has approached near, a note is heard, and the place over which the birds are scattered becomes alive with moving forms and beating wings, and gathering close together they circle swiftly around the observer, or betake themselves to some distant field.

In the far north, on the Alaskan coast, this species is very abundant in the breeding season, arriving in May. Its range is difficult to determine, as it is confused with that of its Asiatic relative, the Pacific Golden Plover, and from which it is very unsatisfactorily separated. It is spread over the eastern shore of Behring Sea, and the Arctic, very common at Point Barrow, and reaches the shores of Norton Sound the latter part of May. They are then in full breeding dress, and present a beautiful sight, flying over or feeding on the flats. The male is brighter in plumage than the female, and at this season is accustomed to utter a sweet, melodious song, most frequently heard during the brief hours of the Arctic night. Nelson describes this as represented by the syllables *tē-lē-lē, tū lē lē wĭt, wĭt wĭt, wē-ū-wĭt, chē lē ū tōō lē-ē*. The three last syllables serve as a call note, he says, but the full song is more often repeated during the night than in the day. The nest is arranged in a circular form among the moss or grass, lined with the latter and dead leaves. The eggs are pale yellowish, with dark reddish brown spots covering the shell, averaging in size 1.90 by 1.30 inches. There is much variation in the color, both in the spots and in that of the shell. The usual number in a nest is four. In its breeding dress, together with its gentle, unsuspicious nature and melodious voice, the Golden Plover is one of the handsomest and most attractive members of the family. By the last of September the southern migration begins, and all have gone by the middle of October. In the Anderson River region, and the Barren Grounds up to the Arctic Ocean, this species also breeds in great numbers, and many nests and eggs were found by MacFarlane. On the California coast the Golden Plover appears to be rather rare, the bulk of the birds passing much farther to the eastward. On the western coast of South America it is an occasional visitor in Chili, but in the Argentine Republic it is present in enormous numbers, less plentiful at the present day, however, than formerly. When congregating, as is their habit, in some marshy place on the pampas during the middle of the day, they blacken the ground to the extent of several acres, and the din of their voices resembles the "roar of a cataract." Hudson mentions one habit they have on the pampas, which I have never witnessed myself, when, a few birds passing overhead catch sight of others on the ground, they descend rapidly and almost vertically on fixed wings to the earth, producing a loud sound like the blowing of a horn. The advanced guard arrives from the north the last of August, and the return journey is begun in March.

## *CHARADRIUS DOMINICUS.*

*Habitat.*—All North America from the Arctic Sea, migrating in winter through South America to Chili and Patagonia. Breeding in the Arctic region from Greenland to Alaska.

*Adult in Summer.*—Resembles so closely the previous species, that an extended description is quite unnecessary. The only appreciable difference is to be found in the under wing-coverts and axillary plumes, which are smoky gray instead of white. Length, about 10 inches; wing, average, 7.09; culmen, .92; tarsus, 1.70; middle toe, .90.

*Winter Plumage.*—Differs from the summer in the under parts, which are white, mottled with brown on neck and breast, sometimes mixed with black.

## PACIFIC GOLDEN PLOVER.

IT is exceedingly difficult to distinguish the Pacific Golden Plover from the American, the only difference being its smaller size and more golden hue. It occurs on the Alaskan coast from the peninsula to Point Barrow. On this stretch of shore line to the island of St. Lawrence, as stated by Nelson, the American Golden Plover is the predominating form, but the specimens grade regularly into the present subspecies. In the interior of Alaska it would appear that the first-mentioned bird is the only one found. The Pacific Golden Plover, or the Asiatic form of the American bird, occurs on the islands of Behring Sea from the Fur Seal to St. Lawrence Island, and Murdoch states that all the Golden Plover obtained at Point Barrow belonged to the American species. The range of this bird on the Alaskan coast cannot be said to have been satisfactorily settled as yet. Its habits are like those of the American Golden Plover.

### *CHARADRIUS DOMINICUS FULVUS.*

*Habitat.*—Asia and islands in Pacific Ocean, Prybilof Island and coast of Alaska. In winter to India, China, and Australia. Breeds in Eastern Siberia from the valley of the Yenisei to the Pacific.

*Adult.*—Almost identical with the American Golden Plover, but more golden above and slightly smaller, the average of twenty-four specimens being, wing, 6.40; culmen, .92; tarsus, 1.72; middle toe, .90.

56. Killdeer Plover.

## KILLDEER PLOVER.

ONE of the most beautiful of all the species, the Killdeer Plover, or, as it is frequently called, Killdee, as an article of food is practically worthless. It is distributed generally throughout the limits of the United States, and while not uncommon along the coast, it is more numerous in the interior. It passes nearly all its time upon the ground, walks and runs with ease and considerable grace, and is constantly in motion, uttering its plaintive cry, which resembles the syllables that form its trivial name. It likes to linger around pools and the banks of streams, and feeds upon worms, insects, larvæ, and small crustacea, and is often seen running over plowed ground in search of whatever insects may have been disclosed in the upturned soil. While usually rather tame and gentle, it nevertheless resents man's appearance on its territory, and continually utters its complaining note, running before him, stopping occasionally to take observations or flying short distances. When on the wing it is a beautiful object, the clear, harmonious-contrasting colors of its plumage making it very attractive to watch, as on firm wings it circles around in easy flight. In the autumn it often is most numerous near the seashore, but I do not remember ever to have seen it actually on the beach. Walking quietly over the meadows or fields, thinking nothing of birds and none being in sight, one is often startled by this Plover rising suddenly from almost beneath his feet, with frequent repetitions of its shrill cry, the last syllable sounded in rapid succession—*dee, dee dee*

*dee*—as though it had no time in its excitement to utter the full sound, *kill-dee*. At such times it flies often in an erratic course for quite a distance, and low over the ground, as if to entice its disturber to follow it, and acts as if its nest was near, although the breeding season may have long since passed. It is a noisy bird, and serves on many occasions as a sentinel, and gives the alarm to other species not so watchful of approaching danger. On this account it is not looked upon with favor by sportsmen who may be endeavoring with well-executed whistling to lure other waders to their place of concealment. Like the Golden Plover and others of the tribe, it frequently stands motionless watching the object of its suspicions, and then running quietly away or rising with shrill cries, informing every other bird within hearing that it is time to be off from that particular locality. Frequently the Killdeer remains all winter in some of the Middle States if the weather is not too severe, but when migrating it travels chiefly at night, often at a great height, announcing its presence by its clear, plaintive note sounded amidst the stars. It breeds in different parts of the land from April to June, and the nest is merely a depression in the ground, lined sometimes with grass. The eggs, four in number, are much pointed at one end, of a cream color, spotted thickly with blackish brown. Sometimes the ground color is a brownish drab and the spots rather small. They measure 1½ inches long by 1⅛ broad.

## *ÆGIALITES VOCIFERA.*

*Habitat.*—Temperate North America from the Saskatchewan to Bermudas, West Indies; Central America to Colombia in winter. Breeding in the proper season wherever found.

*Adult.*—Top of head, nape, back, and wings, grayish brown, some of the

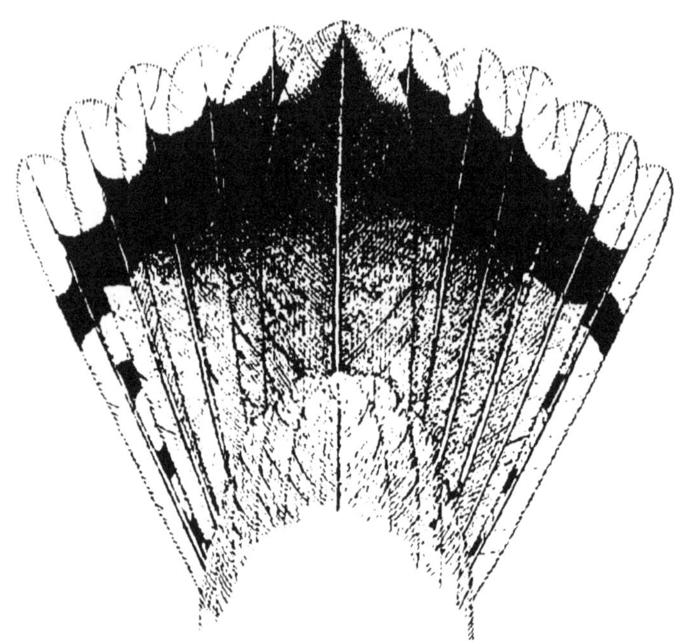

57. Tail of the Killdeer Plover.

feathers on the back margined with rufous; greater wing-coverts broadly tipped with white, forming a conspicuous bar across the wing; primaries, black, inner one with lengthened patch of white on outer webs; rump and upper tail-coverts, ochraceous, darkest on the former; tail, long, the middle pair pale greenish gray, tinged with ochraceous, graduating into black and tipped with buff; rest of feathers, ochraceous, with a subterminal black bar and white tips; forehead, superciliary stripe, and throat, white, extending in the form of a ring around back of neck; stripe from bill to ear-coverts, black; broad band across lower part of throat, and another, narrower, across upper part of breast, jet black; rest of under parts pure white; bill, black; eyelids, bright orange; legs and feet, grayish yellow. Length, 10 inches; wing, 6½; tail, 3½; tarsus, 1⅜; culmen, ¾.

*Downy Young.*—Upper parts, grayish brown, reddish on the wings, mottled with black and buff; a line from bill to eye, another across front of head, one at base of occiput, bordering the white nuchal collar, and another below it, and a line down back, black; front and under parts pure white, with a black bar across upper part of breast.

## SEMIPALMATED PLOVER.

THE Semipalmated or Ring-neck Plover is a migrant in the United States, passing northward in April and May and returning again in August. Along the seacoast it is one of the most familiar species, running rapidly over the sand in company with flocks of Sandpipers, searching for minute insects and shellfish washed up or exposed by the tumbling tides. It is a very gentle and unsuspicious bird, and pays but slight attention to man's presence, pursuing its avocations without regard to what is going on near at hand. This species resembles very closely its European relative, *Æ. hiaticula*, or Ring Plover, but is easily distinguished from it and other American species by the extent of web at the base of all the toes. The Ring-neck is distributed over all North America, most numerous, however, on the seacoasts, scattered about in small groups, and frequenting both the sandy beach and the salt meadows and mud flats left bare by the tides. Although in winter it goes into South America, many pass this season in the southern parts of the United States and in the Bahamas. It is a silent species, uttering its sharp note mainly when alarmed; and when desirous of removing itself from any fancied danger, frequently trusts to its legs, instead of taking flight, and running swiftly among the sand dunes, conceals itself behind tufts of grass or any object affording a temporary hiding place. In the interior, through

58. Semipalmated Plover.

which it passes regularly during the migrations, this Plover is found about the margins of ponds, lakes, and along the shores of rivers. Its breeding grounds are in the Arctic regions, quite across the continent from Greenland to Alaska. MacFarlane found many nests in the Anderson River region quite up to the shores of the Arctic Sea, and Dall also observed it as very common at Nulato and the mouth of the Yukon, while down the Alaskan coast to Sitka it is numerous in the summer. The nest is a cavity in the soil, occasionally lined with dead leaves, the eggs, from two to four in number, drab in color, with scattered black spots and blotches, and with an average measurement of 1¼ inches long by about 1 inch broad. It also breeds abundantly in Labrador and in various parts of British North America. According to Nelson, it is found on both shores of Behring Sea, and along the northeastern coast of Asia, while it has been met with by various naturalists on both coasts of South America, as well as in the interior east of the Andes.

### *ÆGIALITIS SEMIPALMATA.*

*Habitat.*—North and South America, Greenland, Bermudas, West Indies. Breeding in both Arctic and sub-Arctic regions. Migrating from North America in winter to Brazil, Peru, and the Galapagos Islands.

*Adult.*—Forehead, spot under the eye, throat, and ring round the neck, pure white; line over the base of bill, lores, line beneath eye, band across the crown, bar across breast encircling the back beneath white ring, jet black; occiput and nape, back, wings, rump, and upper tail-coverts, ashy brown; tips of greater wing-coverts, white, forming bar across wing; primaries, brownish black, with white shafts, and a white streak on outer webs of innermost one; middle tail-feathers, ashy brown, with a subterminal brownish-black bar, tipped with white, the remainder similar, but graduating into the white of the outer feathers; under parts beneath black breast bar, pure white; bill, orange yellow at base, black at tip; legs and feet, flesh color. Length, 7 inches; wing, 4¾; culmen, ½; tarsus, 1, web between outer toes and the middle one reaching to second joint.

*Immature.*—Similar to the adult, but the black replaced by ashy brown, like the upper parts; maxilla, black · base of mandible, pale orange.

59. European Ring Plover.

# EUROPEAN RING PLOVER.

IN their habits and economy this species and the previous one do not appear to present any especial difference worthy of note. It is included in our fauna from the fact of its breeding on the American side of Davis Bay, on the shores of the Cumberland Gulf.

### *ÆGIALITIS HIATICULA.*

*Habitat.*—Northern portions of Old World from the British islands as far east in Asia as the Taimyr Peninsula. Breeding in Cumberland Bay, Davis Strait, Greenland, Iceland, Spitzbergen, and Nova Zembla also in western Siberia and Turkestan. Winters in Africa.

*Adult.*—Resembles very closely the Semipalmated Plover, but is larger. The only difference observable between the species is, that the present has a conspicuous white spot behind the eye, and the basal web between the outer and middle toe only reaches to the first joint.

## LITTLE RING PLOVER.

THIS is another European species, very doubtfully included in the North American fauna, having even much less grounds for its reception than the preceding ones.

### *ÆGIALITIS DUBIA.*

*Habitat.*—Northern portion of Old World as far east as China. Winters in Africa. Breeds in north of Europe and Asia. Accidental on coast of Alaska and California.

The Little Ring Plover is almost an exact counterpart of the European Ring Plover, but is smaller than the Semipalmated Plover. The differences claimed for it beside its size are, the white on primaries confined to the shaft; base of mandible only yellow; legs and feet dull yellow instead of orange yellow; orbits yellow and iris dark hazel. Length, 6 inches; wing, 4 1/3; tail, 2 1/3; culmen, 5/8; tarsus, 7/8.

60. Little Ring Plover.

61. Piping Plover.

## PIPING PLOVER.

IN its habits the Piping Plover does not differ from those of kindred species dwelling upon the borders of the sea, and obtaining nourishment from the sands washed by the flowing tides. From many of its resorts along the Atlantic Coast, where in former days it was most abundant, it has been driven by the advance of fashion and the influx of the summer's passing population, until it is now found chiefly on the more retired parts of the coast where it is most free from molestation. Although perhaps of not so confiding a disposition as the Semipalmated Plover, it yet can not be regarded as a wild species, though its acquaintance with man has caused it to be at the present time, in most places where it is found, a rather wary bird. It loves to resort to the sandy beaches, where, close to the water's edge, it follows the retiring waves, picking up with great rapidity the insects and small crustacea disclosed upon the sand. Its movements are very quick, and it runs with great swiftness, avoiding with surpassing agility the rush of the incoming wave, that would seem certain to engulf it. At times it skims over the ocean at a short distance from the land, rising and falling over the rolling waves at just a sufficient height to escape their curling crests, or with rapid wing-beats it darts along the shore from one part of the beach from which it has been disturbed to another but a short distance beyond. Sometimes its brief flight is made directly over the sands, again by a lengthened curve over the water. When the tide is high, in company with other beach birds, this Plover retires to the dunes and sandy districts at the back of the seabeach, and rests and sleeps away the hours, if unmolested, until the retiring tide again lays bare the places from which its food is gathered. This consists of worms, insects of various kinds, and small crustacea, in the pursuit of which it is very diligent. The note of the Piping Plover is soft and musical, and is frequently uttered when startled, or as the bird flies along the beach. The nest is but a depression in the sand, extremely difficult to find, as the eggs resemble so much their usual surroundings as to be almost imperceptible to the eye. Their color is a light yellowish drab, spotted with black or blackish brown, measuring 1¼ inches in length by 1 in breadth. The mother employs all the well-known artifices to draw away the intruder from the vicinity of the nest, such as lameness, inability to fly, etc., and the young run as soon as they leave the egg, and are great adepts at hiding, squatting, and remaining motionless, until almost stepped upon. Their downy plumage so assimilates the chicks to the sand around them that unless they discover themselves by moving, it requires a very keen eye indeed to distinguish them from the numberless tufts dotted about the higher portions of the beach. Although so essentially a "beach bird," this Plover is by no means unknown in the interior, and is found in considerable numbers around the shores of the Great Lakes, as well as others, as far west as Wisconsin. It is a migrant in Manitoba, and although replaced in the Missouri region by a race very like it in appearance, it is doubtless also found with members of the Belted Piping Plover in the Mississippi Valley, even to Texas.

It breeds apparently throughout its range, from Cuba to the Magdalen Islands, perhaps even in Southern Labrador, but more generally from Virginia northward. As a rule the Beach Bird, as it is called in many places, prefers to trust to its legs rather than to its wings, but is able to perform long journeys in a brief period, and during its migrations proceeds at a considerable elevation. When fat the flesh of this bird is very palatable, though at times it has a sedgy flavor.

### *ÆGIALITIS MELODA.*

*Habitat.*—Eastern North America from Labrador along the Atlantic Coast, west to the Great Lakes. In winter to Bermuda, Cuba, and West Indies, Breeding most commonly from Virginia northward.

*Adult Male.*—Band across forehead between eyes, and another around back of neck, and on sides of breast, jet black; forehead, ring around neck above the black, and entire under parts, pure white; head on top, ear-coverts, back, and wings, pale ashy, with a brown tinge; rump and upper tail-coverts, white, washed with ashy; primaries, dark brown; shafts and great part of inner webs, white, inner primaries having outer webs also mostly white; tail, white at base, graduating on all but two outer white feathers into a subterminal black band, and with white tips; bill, orange; tip, black; legs and feet, orange yellow. Length, 7 inches; wing, 4½; culmen, ½; tarsus, ⅞; middle toe, ¾.

*Female.*—Similar in plumage to the male, but with the black bars more of a brownish hue and less in amount.

*Young.*—Without the black band, and the nuchal collar ashy brown.

## BELTED PIPING PLOVER.

IN its habits this race does not differ from the eastern Piping Plover. It breeds in various portions of its range, and possibly as far eastward as Lake Koshkonong, in Wisconsin, where, from the behavior of the birds, Mr. Nelson supposed it nested on the beach, but he found no nest or eggs, although he procured one from a female shot at Waukegan. It occasionally occurs on the Atlantic Coast among flocks of the Piping Plover, straying from its home in the interior, in the same way as the eastern bird appears among the western race. In Texas, in the vicinity of Corpus Christi, it appears to be quite common.

*ÆGIALITIS MELODA CIRCUMCINCTA.*

*Habitat.*—Mississippi Valley north to Lake Winnipeg. Occasional on Atlantic Coast. Breeding in its range.

*Adult Male.*—Resembles the plumage of the Piping Plover, but has the black on the breast continuous, forming an uninterrupted band. Young birds do not have this, and in individuals among adults it varies in intensity, caused possibly by difference of sex.

62. Belted Piping Plover.

63. Snowy Plover.

## SNOWY PLOVER.

THE Snowy Plover is a Western bird, occurring from the Great Salt Lake to the Pacific. It is abundant at times on the shores of the Great Salt Lake, and may possibly breed there. On the coast of California, especially in the southern portion, it is very common, dwelling upon the shore, and having all the habits of the Piping Plover, following the waves as they recede upon the beach and running rapidly back to escape those that come rolling in. It lays its eggs in a hollow in the sand; but once Mr. Henshaw found them deposited on a glittering collection of bits of mother-of-pearl and broken shells. They are of a light clay color, spotted with black, measuring 1¼ inches in length by about 7/8 inch in breadth.

The Snowy Plover goes in small flocks, and is a very busy little bird, running nimbly over the sand intent upon its diet of insects and crustacea, for which it is always engaged in active search whenever the tide permits.

It has a low, rather mournful note resembling that of the Piping Plover, which it utters frequently when any one intrudes in the vicinity of the eggs or young; the female at the same time, by all the usual artifices and mournful pleadings, endeavors to entice the observer in pursuit of herself and away from her treasures.

In its southern migration it is found on both coasts of Central America, and in the United States east of the Rocky Mountains; has been obtained in Kansas and the Indian Territory; also it is found in Texas.

## *ÆGIALITIS NIVOSA.*

*Habitat.*—Western North America from Great Salt Lake to the Pacific, north to Cape Mendocino. In winter through Central America along the western coast of South America to Chili; Western Cuba. Breeding in its northern range.

*Adult Male.*—Forehead, superciliaries, indistinct collar on back of neck, and entire under parts, pure white; band across front of crown (in some specimens a broken line across lores), ear-coverts, and broad patch on either side of the breast, jet black; crown and nape, reddish buff; upper parts, grayish brown; primaries, blackish brown, with white shafts; inner secondaries, broadly marked with white; middle tail-feathers, dusky brown, growing paler on the others to the outermost pair, which are pure white; bill, black; legs and feet, yellowish in skin. Length, 6¼ inches; wing, 4¼; culmen, ⅝; tarsus, 1; middle toe, ½ inch.

*Young.*—Black markings replaced by ashy brown, feathers of the back faintly margined with brownish white.

The black line across the lores from the bill is often, probably in the vast majority of specimens, obsolete; again it is quite distinct in some males in the breeding plumage, but rarely is it perfect as in the European *Æ. contiana* or Kentish Plover, to which the present species bears a strong resemblance.

64. Mongolian Plover.

## MONGOLIAN PLOVER.

A VERY handsome bird, the Mongolian Plover is an Asiatic species and owes its appearance in the American fauna from the fact that the captain of the English ship Plover obtained two specimens on Choris Peninsula, in Kotzebue Sound, in 1849. They were undoubtedly stragglers from the islands in Behring Sea, on some of which the species is quite numerous. In the Commander Islands it is a common summer resident, one of the handsomest of the beach birds, running with great rapidity over the pebbly shore in search of food. It breeds in June, the nest, as described by Stegneger, was placed in a slight hollow in the ground between the stems of four *Angelica archangelica* and formed of leaves, stems, and seeds of the same plant. It was about forty feet from high-water mark. The eggs resembled those of the Semipalmated Plover, but larger and of a deeper color, and were three in number.

### *ÆGIALITIS MONGOLA.*

*Habitat.*—Northern Asia; in winter from the Red Sea to the Malay Archipelago and Australia. Accidental on Choris Peninsula, Alaska. Breeds from Eastern Turkestan to the valley of the Amoor.

*Adult Male.*—Forehead, white, bordered by black, formed by a line from bill to eye, and another across front of crown; line beneath the eye, a continuation of loral stripe, and ear-coverts, black; stripe from behind the eye, buff, graduating into bright rufous; crown and nape, brownish gray, mottled with rufous on the anterior portion; a narrow nuchal collar, widening into a broad band that crosses the breast, bright cinnamon rufous; upper parts and wings, brownish gray, with a slight greenish tinge, lighter on the rump; tips of greater wing-coverts, white, forming a bar; primaries, blackish brown, first with white shaft, remainder with only apical half white; upper tail-coverts, pale grayish brown in the center, lateral ones white; tail, dark brownish gray on central pair, graduating in the rest to the pure white of the outer feather, and all except middle pair tipped with white; entire under parts, excepting breast band, pure white; bill, black; legs and feet, gray, tinged with olive. Length, 6¾ inches; wing, 5½; culmen, ¾; tarsus, 1⅛; middle toe, ¾.

*In the Winter Plumage* the breast is crossed by a narrow grayish-brown band instead of cinnamon rufous, and the black markings of the head are replaced by grayish brown.

65. Wilson's Plover.

## WILSON'S PLOVER.

THIS species, which in its plumage so much resembles the Semipalmated Plover, is more a sojourner of the southern portion of our seacoast, and is not common much above the sand-beaches of Virginia. It is true it is found on the shores of Long Island and occasionally proceeds onward as far as Nova Scotia, but it is only a straggler there, and its more congenial home is on the southern Atlantic seaboard. On the Pacific it is found on the California coast, very abundant in the southern part, and then it goes on both coasts of South America to Brazil and Peru. It is a beach bird, fond of running over the sand, and seeking its food along the margin of the seething, tumbling waves hurled on to the sounding shores from ocean's heaving breast. Like the other Plovers and various Sandpipers with which it often associates, it is active in avoiding the rush of the water as it is flung from the breaking waves, and with surprising agility snatches from side to side any toothsome morsel, such as insect or small shellfish washed about by the curling water, in which it may often be standing half-breast high. Wilson's Plover migrates in small flocks, arriving in April or May, and soon the little company breaks up into pairs, and the important business of nesting with its attendant joys and cares commences. Back from the beach where the grass is short a hollow is scooped in the ground so shallow that it would easily be passed unnoticed, or sometimes in a scanty tuft of grass, usually three eggs are deposited, sometimes four, pale olive drab in color, spotted and splashed profusely with blackish brown, 1.40 inches long by 1 inch broad. Terns and Plovers sometimes select the same locality for a breeding ground, and then the air is filled with active, graceful figures wheeling and flying about in all manner of beautiful curves and evolutions; while the ear of the observer is made to ring from the various cries uttered in never-ceasing chorus by the loving, excited creatures guarding their treasure, and the males encourage the patient females with low melodious notes, or with frantic scream and downward swoop, both sexes strive to terrify and drive away some intruder on their domain. The usual note is a "whistle and chirp," a mixture of both, very different from that of allied Plovers. In the more southern seaboard States this species is quite abundant, and in flocks of two or three dozen present a most attractive sight as they run about the beach, or with steady easy flight pass rapidly from one point to another, uttering, as they speed along, their peculiar low note. Wilson's Plover is easily recognizable among the American species by its large, strong bill, much larger in proportion to the size of the bird than that of any other Plover.

## *ÆGIALITIS WILSONIA.*

*Habitat.*—Coasts of North America from Long Island, and Lower California, southward to West Indies, Brazil, and Peru. Occasional in Nova Scotia.

*Adult Male.*—Forehead, line over eye, white; lores, brownish black; black band across front of crown; crown, nape, and ear-coverts, brownish gray, tinged with buff; throat and upper part of neck, and a band around back of neck nearly meeting in the center, pure white; back, wings, and rump, brownish gray; lesser wing-coverts, margined with whitish; greater coverts, tipped with white, forming a bar; primaries, dark brown, with white shafts; middle upper tail-coverts, brownish gray; lateral ones, white; a broad black band across upper part of breast; rest of under parts, pure white; bill, black; legs and feet, flesh color. Length, 7½ inches; wing, 4¾; culmen, ¾; tarsus, 1¼; middle toe, ¾.

*Female.*—Like the male, but the black markings replaced by brownish gray tinged with rusty, the breast band tinged with buff.

## MOUNTAIN PLOVER.

THIS is a prairie Plover, never resorting to the beach, but dwelling upon the plains away from the water, preferring the grassy districts, sometimes being found even in sterile tracts covered with the sage-bush and kindred plants. The Mountain Plover goes in quite large flocks, is naturally tame and unsuspicious, permitting intrusion upon its haunts without evincing any especial alarm, though after having made man's acquaintance, and conscious of the danger there is in his society, it becomes shy and wary and will not allow a near approach. It rises from the ground by several quick flaps of the wings, and wheels and circles over the prairie in beautiful evolutions, exhibiting alternately the under and upper sides of the body in the manner of many Sandpipers; the dark back is brought into strong contrast with the white under parts, as the latter flash in the sunlight, when brought into view of the spectator. This species runs rapidly and easily, with lowered head, and after proceeding for a short distance, stops abruptly and remains motionless, apparently to survey the ground about it or to observe the cause of its temporary alarm. The flight is usually performed low over the ground, flapping and sailing, with decurved wings, and it runs a few steps after alighting; and if much alarmed, the bird will squat close to the ground, trying to conceal itself as much as possible. In the desert region of New Mexico this bird is at times very numerous, and also in Arizona and

66. Mountain Plover.

Southern California. The note is usually a low whistle, changing to a shrill, loud utterance when alarmed. The Mountain Plover feeds chiefly, if not altogether, upon insects, such as grasshoppers, crickets, beetles, ticks, and possibly worms when available, and is often very fat and in fine condition. In the breeding season, when the birds are scattered about the plains in pairs, they are usually silent, as if desirous of drawing as little attention to themselves and their important occupation as possible. The nest, which is merely the usual depression in the ground, perhaps lined with a little grass, contains generally three eggs, which are of an olive or brownish drab, spotted all over with blackish brown. The spots are small and most numerous at the larger end. After the young are hatched, which occurs in June or July, according to locality, sometimes several broods and their parents associate together in small companies. This Plover is not recorded north of the United States. Its nearest relative is the Asiatic Plover, and differs from the majority of the members of the genus in the lack of black bands on the chest, which is peculiar to other members of the group, making them rather conspicuous objects in the family.

## *ÆGIALITES MONTANA.*

*Habitat.*—Western United States from the Great Plains to the Pacific. Accidental in Florida. Breeding in Dakota, on the Platte River, and other points of its range.

*Adult in Breeding Dress.*—Front and stripe over the eye, white; fore part of crown and stripe from bill to eye, black; rest of crown and entire upper parts, light grayish brown, sometimes the feathers edged with reddish buff; primaries, brownish black, lightest on inner webs, and with white shafts; upper tail-coverts, pale brown; tail, dusky brown, all the feathers, save the outermost pair, blackish subterminally, and tipped with white; entire under parts, dull white; across the breast an indistinct ochraceous bar, darkest on the sides; bill, black; feet and tarsi, orange yellow. Length, $8\frac{1}{2}$ inches; wing, 6; culmen, $\frac{7}{8}$; tarsus, $1\frac{1}{2}$; middle toe, $\frac{3}{4}$.

*Winter.*—The black markings on the head are absent, and the plumage more tinged with buff.

*Young.*—Head, neck, and upper part of breast, grayish brown, like the back; under parts tinged with buff.

67. Plover-billed Turnstone.

## PLOVER-BILLED TURNSTONE.

SURF BIRD, as it is usually called, or the Plover-billed Turnstone, is rare on our shores, although it ranges nearly the entire length of the western seaboard of the two Americas. But little is known of its habits, and its breeding place and eggs have not yet been discovered, but it is probable that it nests in the northern part of its dispersion and winters in the lower part of South America. In Alaska it has been taken near Sitka, and also on the island of St. Michael's, and probably it frequents the outlying islands and capes in Behring Strait and Sea. At St. Michael's the specimens obtained were on the muddy flats ; and at the mouth of the Columbia River it has been seen on the rocks near to the sea, where the spray from the heavy surf flew over it as it searched for its food. Its favorite haunts are said to be the same as those of the Wandering Tattler, but little is really known about it.

It is said to visit the Sandwich Islands, but there is no authentic record of its ever having been taken in that group.

*APHRIZA VIRGATA.*

*Habitat.*—Pacific Coast of North and South America from Alaska to Chili. Breeding place unknown.

*Adult in Summer.*—Head, neck, back, and breast, mottled with black and white, the former, in shape of streaks on head and neck, but of crescentic bars, on back and breast ; scapulars, black, with large spots of rufous ; wing-coverts, dark grayish brown ; tips of greater coverts, white, forming a bar across the wing ; primaries, blackish brown ; shafts, white ; rump, brownish black, feathers indistinctly margined with white; upper tail-coverts, pure white; tail, white on basal half, remaining portion black, narrowly tipped with white; flanks, abdomen, and lower tail-coverts, pure white, with black spots on flanks and tail-coverts; bill, black; base of mandible, grayish yellow; legs and feet, gamboge yellow. Length, 10 inches; wing, 7; culmen 1; tarsus, 1¼, middle toe, 9-10.

*Winter Plumage.*—Head, neck, breast, and upper parts, dusky brown or slate color; rest as in summer.

68. Turnstone.

# TURNSTONE.*

THERE is much individual variation among examples of this species, caused mainly by sex and season, and while birds in the plumages described hereafter bear a very general resemblance to each other, still certain difference will be seen that renders even the most minute description of any particular individual not applicable in all respects to any other, but the species is so widely distributed and generally known, that it is not likely to be confounded with any other, even by one whose knowledge of the waders is limited.

The common Turnstone is generally distributed throughout North America, most numerous along the seacoasts of both oceans, and is found in the interior of the continent on the shores of the Great Lakes, as well as other bodies of water of less dimensions, and also the shores of streams, and is known by the popular names of Brant Bird, Horsefoot or Horsefoot Snipe, Beach Bird, Streaked Back, Calico Bird, Checkered Snipe, and many others. It does not associate in large flocks, but goes in little companies, often singly, and wanders over the beach in search of insects and small crustacea. It also feeds upon the spawn of the king crab, or "horsefoot," as it is usually called, and from this takes one of its trivial names. In searching for its food, it rolls over the small pebbles lying everywhere upon the beach, and examines earnestly the exposed spot for any insect its action has disclosed. The various small companies pass northward in April to their far northern breeding grounds, returning in August, and linger frequently on the beaches of the Middle States until quite late in the autumn, and in some of the Gulf States individuals remain during the entire winter. Sometimes the Turnstone will alight on the branches of mangroves or on stumps standing in the water, and if unmolested will remain for a long time in such situations. The bright plumage of these little birds makes them very attractive objects as they run about the sands, the sunlight glancing upon their variegated-colored backs, and when on the wing the strong contrasting hues of its plumage appear to the greatest advantage. Its flight is fast, performed usually at a low elevation ; first a few rapid beats, then sailing along on motionless pinions, succeeded by some more quick flaps. It is a very energetic and persistent bird, very active, runs with great speed, and stopping constantly to investigate some object that has caught its eye or some spot likely to yield some favorite morsel. When it finds a stone rather difficult to move from its weight, it tugs at it with all its might, not infrequently pushing against it with its breast ; and if the object is too deeply implanted in the sand to be turned over in the ordinary way, it endeavors to undermine it and roll it over in the hole thus made. As a rule, the Turnstone is not a shy bird, but at the same time does not permit an observer to approach very near, but rises, uttering a few low notes, and moves on for a short distance. It has a clear, loud whistle, composed of one or two notes, sometimes the single one being repeated, forming three, uttered rapidly. It is a good swimmer, and not at all incommoded upon the water, sometimes deliberately seating itself upon the surface, where it moves about with ease and grace. It has the usual habit witnessed in beach birds, of running a few steps, then stopping suddenly and remaining motionless, as though in deep thought, or to take especial observation of some particular object, and then running swiftly on again, its course abruptly terminated by some stone lying in its path, and which it is obliged to turn over and discover what may be hidden beneath. Bold and rocky shores are as much frequented by the Turnstones as the sandy beaches, and it busies itself with the shells clinging to the stones or grass cast up by the sea. As far as man has penetrated into the frozen regions of the north, the Turnstone has been found breeding quite across the North American continent. It arrives at its nesting place the last of May, and is quite noisy, as it flies from one feeding place to another, when disturbed. On the Alaskan coast it breeds in the same localities and mingles with the Black Turnstone, the two species flying together in small flocks. The nest is the usual depression in the ground, lined with a few withered leaves, and the eggs, four in number, vary greatly in their coloration, from a pale olive green to buff, spotted and blotched with dark or blackish brown and purplish gray, and measure 1.60 to 1.72 inches in length and from 1.13 to 1.23 inches in breadth. When disturbed on its breeding ground the Turnstone becomes very excited, running and flying about, uttering shrill notes. The parents lead the young to the shore soon after they are hatched, which happens toward the middle of July, and in August they commence their long journey to their winter quarters in far southern lands. In one of the South Pacific Islands (Nawado or Pleasant Island, latitude 0° 25′ south, 167° 5′ east longitude) the natives keep this species in small cup-shaped cages, and employ them in fighting, one bird against the other, in the manner of game-cocks. It is probable that captivity or the effects of food given them produces this pugnacity, a trait not exhibted by the Turnstone in its wild state, as any species of shore bird would seem to be the very last one to select for the purpose of combat in a prize-ring.

### *ARENARIA INTERPRES.*

*Habitat.*—Cosmopolitan; found in nearly every part of the world; almost exclusively a shore bird. It has been met with in Central Asia and Africa, in North and South America from Alaska to Straits of Magellan, and interior of North America. Breeds in circumpolar regions, on shores of Arctic Sea.

*Adult in Summer.*—Head, a large spot on the lores, top of head, ear-coverts, nape, back of neck, and upper part of back, extending to sides of breast, chin, and throat, white, streaked on crown with black; rest of head, sides, and front of neck, and upper part of breast, jet black; back and scapulars, black, center of mantle and scapulars, blotched with rufous; lesser wing-coverts, rufous; greater coverts, black, margined broadly with white, forming a conspicuous bar across the wing; primaries, dark brown, tipped with white, and with white shafts; lower part of back and rump, white; middle upper tail-coverts, black, margined with rufous, lower ones white; tail, white, with broad subterminal black band and tipped with white; under parts below, black; breast, pure white; bill, black; feet and legs, orange red. Length, 9 inches; wing, 6; culmen, ⅞; tarsus, 1; middle toe, ¾.

*Winter Plumage.*—Resembles the summer dress, but has less rufous, and the black feathers on head, neck, and breast have white margins.

*Young.*—Upper parts, dark brown, mottled with black and pale brown, and some rufous on scapulars; top and sides of head, pale brown, streaked narrowly with black; breast, mottled with black and pale brown; throat and under parts, pure white.

69. Black Turnstone.

## BLACK TURNSTONE.

THE Black Turnstone is only found in North America on the Pacific Coast, and in some localities is quite numerous. In its habits it resembles the better-known species that roams around the world, but, unlike its relative, is not possessed of the roving spirit that carries that bird into nearly every known land. It is most abundant in the northern part of its range, visiting Southern California in restricted numbers, at times in the company of the common Turnstone. It is plentiful on the coast of Behring Sea and on the shores of Alaska, near Point Barrow, as well as on various islands. It visits the wet flats and the shores of brackish ponds, also marshy places, and in all such localities it breeds.

The note is a piping cry, resembling *weet, weet, too-weet* in the breeding season, and when disturbed near the nest utters a sharp *peet, weet, weet*, resembling that of the Spotted Sandpiper. The species reaches its nesting grounds in May, and breeds wherever it happens to stay. The young are on the wing in July, and resort to the coast to seek their food along the margin of the sea. The Black Turnstone displays the same distress as the other species when its breeding place is invaded, circling about and never a moment at rest.

The eggs are an olive drab, profusely covered with dark spots, similar to those of the common Turnstone.

By the middle of September all have left their northern homes for southern lands.

### *ARENARIA MELANOCEPHALA.*

*Habitat.*—Pacific Coast from Alaska to Santa Margarita Island, Lower California. Breeding from Alaska to British Columbia. Accidental in India.

*Adult in Breeding Plumage.*—Spot in front and behind the eye, and streaks on forehead and ear-coverts, white; rest of head, neck, back, and scapulars, brownish black, with greenish reflections, edged on scapulars and greater wing-coverts with white, forming a bar on the wing; primaries, dark brown on outer webs, whitish on inner, and with white shafts; lower part of back and rump, pure white; short upper tail-coverts, black, longest and lateral ones white; tail, with a very broad subterminal black band; throat and chest, blackish brown, lighter than the back, feathers edged with light brown and with white streaks on breast; rest of under parts, pure white; bill, black; legs and feet, greenish yellow. Length, 9 inches; wing, 6; culmen, 1; tarsus, 1; middle toe, ⅞.

*Winter Plumage.*—Like the summer, no white on head or neck in front.

70. European Oyster-catcher.

## EUROPEAN OYSTER-CATCHER.

ITS only claim to a place in the North American fauna is based on the fact that the European Oyster-catcher has appeared in Greenland. In general appearance it closely resembles the common American species distributed generally throughout our country, but is a smaller bird, and may be at once distinguished by its white rump, the American species having this part dark brown in the center. It is common in the British Islands, found also along the shores of the Atlantic throughout the eastern coast of north-western Europe, and is met with on the banks of the great rivers in eastern Europe and western Asia. It is replaced in eastern China and Japan by a closely allied species or race, with less white on the wings; by some ornithologists, however, this is not deemed a mark of sufficient importance to have even a subspecific value. In winter it ranges into Africa, as far south on the east as Mozambique, and on the west to Senegambia. In its habits and economy it resembles the American Oyster-catcher, scratches a hollow near the seashore in the gravel or amid the stones, placing at the bottom a few weeds, on which three, sometimes four, eggs are deposited. These are buff in color, spotted with blackish brown and purplish gray, measuring about 2 1-5 by 1½ inches.

*HÆMATOPUS OSTRALEGUS.*

*Habitat.*—Eastern hemisphere from the Atlantic to the valley of the Obb. Frequenting the seacoasts to Archangel, thence eastward on the banks of great rivers. Occasional in Greenland. Breeding in Eastern Siberia and Kamtshatka.

*Adult.*—Head, neck all around, and upper parts of breast, black; back and wings, brownish black; primaries, brownish black on outer webs, white on great part of inner, and extending on to outer webs on the fourth and fifth; shafts, white; greater wing-coverts, rump, upper tail-coverts, and entire lower parts, pure white; tail, white on basal half, remainder brownish black; bill, orange vermilion; legs and feet, dull crimson; iris, crimson. Length, about 16 inches; wing, 9¼; culmen, 2⅞; tarsus, 2.

71. American Oyster-catcher.

## AMERICAN OYSTER-CATCHER.

THE American Oyster-catcher is more a resident of the shores of the South Atlantic States, and only an occasional visitor to the coast north of New Jersey. Formerly it was more abundant at the north, but at the present day on the beaches of Long Island it has become a rather rare bird, not often seen among the great flocks of waders that pass regularly along its seaboard in their migratory journeys. At all times it is a very shy and wary bird, exceedingly watchful and permitting no one to approach, but keeping itself at a safe distance from all known danger. It haunts the seabeach, where it runs with great swiftness or walks in a stately manner, striking its powerful bill into the soil or sand in search of insects or fiddler crabs, on which it usually feeds. If an attempt be made to draw near to it when thus engaged, it ceases its occupation for a moment, and then, with a loud shrill cry, spreads its wings and flies off, generally going a long distance before stopping, frequently altogether out of sight. It is rather a solitary bird, going in pairs or in small companies, and is rarely seen away from the beach, in this respect being entirely different from its European relative, which frequents the banks of rivers in the interior. On the coast of Massachusetts and farther northward it can only be regarded as a straggler. The probable reason for this is the large summer population that now resorts to the seacoast, making its usual haunts liable to too much intrusion to suit so shy and timid a bird. I have never known this species to come to decoys, although possibly it may do so in places where it has not been molested nor learned to fear its great enemy, man; but sometimes it is killed when flying along the beach, as it swings towards decoys set out to entrap its less wary relatives, the Plovers and Sandpipers, which frequent such localities. The majority of the specimens are procured by creeping or stalking them, when the ground is sufficiently favorable to admit of an approach near enough for a shot. They fly with rapid beating of the wings, and execute many and beautiful evolutions in the air, when small companies wheel and circle about actuated as by a common impulse, their large size and the shining white of the bodies and wings rendering them very conspicuous objects, and their well-trained, soldier-like movements affording a most pleasing and attractive sight. The nest is but a hollow in the ground, in which are deposited two or three eggs, cream color, blotched with dark brown, of an oval shape, measuring about 2¼ inches in length by 1⅝ in breadth. The female sits on them chiefly at night, trusting to the heat of the sun and sand to carry on the incubation during the day, as is the custom with other birds which lay their eggs in similar localities. The parents exhibit all the signs of great distress when they consider their treasures, of either eggs or young, in danger, flying about the intruder on their domain, uttering loud cries and endeavoring, with all artifices at their command, to lure him from the spot. The young run as soon as hatched, and are very skillful in concealing themselves when danger approaches, squatting and remaining motionless, and they are so like the sand as to be practically invisible. On Long Island this species is called "Flood Gull," but it is generally known by the common appellation of Oyster-catcher, said to be given from the alleged habit of seizing the oysters when it finds them with the shell partly open. This bird is numerous on both coasts of South America, and has been seen in Patagonia where it was supposed to be breeding.

## *HÆMATOPUS PALLIATUS.*

*Habitat.*—Occasional on Atlantic Coast from Nova Scotia to New Jersey, thence southward to Brazil on the east, and from Lower California to Patagonia on the west; Tres Marias, Bahamas, Cuba, and Galapagos Islands. Breeding in the islands and in most parts of its dispersion on the continent.

*Adult.*—Head, neck, and upper part of breast, uniform black; back, wings, and center of rump, dark brown; greater wing-coverts, and base of secondaries, sides of rump, upper tail-coverts, and entire under parts, pure white; primaries, blackish brown, with purple reflections; tail, white at base, remainder dark brown; bill, vermilion; legs and feet, pale flesh color; iris, yellow. Length, 17 to 21 inches; wing, 10¼; culmen, 3¼; tarsus, 2¼; middle toe, 1¼.

The measurements of individuals vary considerably, and there is much difference often seen in length of bill and wing, but it is always larger than the European bird, and there is no other species except, possibly, *H. frasari*, in North America, with which it can be confounded. The differences between these will be found in the article on Frazar's Oyster-catcher.

# FRAZAR'S OYSTER-CATCHER.

MR. FRAZAR obtained three specimens of this bird to the northward of La Paz, on the Gulf of California, and these comprise the foundation for this species. It was said to be common in the locality and evidently preparing to breed upon the sandy islands and shores of the gulf. It would seem desirable to compare more examples, not only with the common Oyster-catcher, but especially with the Galapagos bird, in order to arrive at an entirely satisfactory decision as to the exact status of this form. The description given below will enable any one who may have specimens from the Gulf of California to ascertain if they agree with those sent to Mr. Brewster. This bird has also been seen on Los Corronados Islands, San Quentin Bay, and Cerros Island; also at Magdalena Bay, where it was common, and on Santa Margarita Island. They mated here in January, were shy, ran rapidly along the beach and took wing, uttering a loud, clear whistle, and after a short flight alighted at the water's edge. They fed upon small bivalves.

*HÆMATOPUS FRASARI.*

*Habitat.*—Western Mexico.

*Adult.*—Similar to both the common Oyster-catcher and the species found in the Galapagos Islands. "Differing from the North American bird in having a stouter and more depressed bill, little or no white on the eyelids; the back, scapulars, and wing coverts richer and deeper brown, the primaries and tail-feathers darker; the upper tail-coverts more or less varied with brown and white; the lateral under tail-coverts marked with brown; the bend of the wing and greater under primary coverts mottled

72. Frazar's Oyster-catcher.

with black and white; from the Galapagos species in the rather shorter bill and distinctly brown (instead of sooty black) back, scapulars, and wing-coverts, dark markings on the under tail-coverts, and greater amount of white on the under primary coverts; from both the above-mentioned species in the broad zone of mottled black and white feathers extending across the breast. Extreme measurements, three specimens, all males: wing, 9.75–10.27; tail, 3.90–4.26; tarsus, 2.18–2.30; bill, length from nostril, 2.35–2.37; from feathers, 2.99–3.05; depth at angle, 49.53."—*Brewster.*

## BLACK OYSTER-CATCHER.

THE Black Oyster-catcher, while met with throughout the Pacific Coast of North America, is most numerous in the north of its range, and is common in Alaska, where it is one of the characteristic birds of the seashore, and is also a summer resident of the Aleutian chain of islands. It appears to breed throughout its dispersion from the far north to Santa Barbara in California. Below our limits on the Pacific it meets, if it ever wanders so far to the south, the allied species or race, *H. ater*, of lower South America. I do not think, however, that there are any records of the two forms having been seen together, as the South American's northern limit is Chili. This bird in many localities exhibits all the shyness shown by other species of the genus, while again it has appeared tame and allowed one to approach quite near to it. It does not frequent sandy beaches so much as the other Oyster-catchers, being more partial to rocky shores, obtaining its food, consisting of different species of mollusks, and crustaceans, from among the seaweed thrown up by the waves. The legs are short and feet powerful, enabling it to scramble easily over the slippery rocks, while the wedge-like bill is an admirable instrument for prying open the small bivalves. This bird has a curious habit of standing on some shelving rock and calling to another a little distance away, who replies to his friend, and this conversation is kept up for a long time, until their peculiar whistle becomes one of the familiar evidences of bird-life in the region it frequents. Its walk is graceful,

73. Black Oyster-catcher.

slow, but somewhat stilted, and as it moves along, it is in the habit of bobbing its head, keeping time with its steps. It arrives in the Aleutian Islands in May, and in June the eggs are laid in a depression in the gravelly beach, perhaps near to some rocky cliff. The eggs are pale olive buff, speckled with brownish black or purplish gray, averaging in size 2 1/6 by 1½ inches. The Black Oyster-catcher is a noisy bird, and sometimes when they are numerous, their cries are sounded continuously, and this habit seems to be indulged most frequently and persistently in the vicinity of their breeding places. In its general habits it resembles very closely its relatives found along the same coast, and it is occasionally seen in company with the common Oyster-catcher, or possibly with the newer form from Lower California. It is a large bird, and its somber, unattractive plumage is relieved, and its appearance rendered more acceptable, by the brightly colored bill and pale legs, affording a strong contrast to its mournful dress.

### *HÆMATOPUS BACHMANI.*

*Habitat.*—Pacific Coast of North America from Alaska, Commander and Kurile Islands to La Paz, Lower California. Breeding throughout its range.

*Adult.*—Head and neck, black; entire rest of plumage, blackish brown; bill, vermilion; legs and feet, pale flesh color; iris, yellow. Length, 17 inches; wing, 9½; culmen, 2¾; tarsus, 1¾; middle toe, 1½.

## MEXICAN JACANA.

THIS Jacana is properly a native of countries lying south of the border of the United States, but numerous examples have been procured in Texas along the valley of the Rio Grande. The first notice of its occurrence within our limits is recorded in the *Bulletin of the Nuttall Ornithological Club for* 1876, when Dr. Merrill saw some near Fort Brown, in Texas, and wounded one, but did not secure it. It is a very active bird, continually on the move, running over the floating leaves of the various water plants that carpet with green the surface of ponds and lakes, raising its wings as if to lessen even its light weight, the long toes spreading out and covering a large part of the great leaf of the lily. The young follow the parents as soon as hatched, and the male assists the female in rearing them. They are very pugnacious in defending their chicks, and also very bold, approaching an intruder quite close, and uttering loud cries of defiance, very comical in so small a bird, although doubtless the spur on the wing could then be employed with very uncomfortable results. These birds are difficult to secure when shot, for usually the plants, on the leaves of which they run about, are in too deep water for any one to wade, and a boat is necessary to reach them. The difference of plumage between young and old is so great that they might easily be regarded as belonging to two distinct species. The nest of the Jacana is flat and formed of grasses interwoven beneath so as to float upon the water. The eggs are dark olive, marked with black or dark-brown blotches and streaks, measuring about 1¼ inches in length by 1 in breadth.

## *JACANA SPINOSA.*

*Habitat.*—Lower Rio Grande Valley, Texas, Mexico, Panama, and Colombia, South America; Cuba, Hayti.

*Adult.*—Head, neck, upper part of back and breast, black, with green and purple reflections; lower back and wings, purplish chestnut; primaries and secondaries, pale yellowish green, bordered on first with blackish brown; rump, upper tail-coverts, and tail, dark purple; lower portion of breast and flanks, dull maroon; abdomen, thighs, and under tail-coverts, dull brownish maroon; frontal leaf or wattle divided into three lobes on top, broad above, narrowing where it joins the forehead, red or orange in life; base of mandible, bluish white, with a space of carmine between it and the wattle; rest of bill, bright yellow; spur on wing, long and sharp at point; legs and feet, dull olive. Total length, 9 inches; wing, 5½; tail, 2¼; bill, 1⅜; tarsus, 2⅛.

*Young.*—Top of head and nape, pale brown, a yellowish white stripe from base of maxilla to nape; black stripe behind the eye, widening as it goes and passing down the side of neck and crossing upper part of back; back and wings, pale bronzy brown; primaries and secondaries, like the adult; rump and upper tail-coverts, purple; tail, purplish black; frontal wattle, rudimentary; chin, throat, sides and front of neck, and entire under parts, white, with a strong buff tinge on the upper part of breast; bill, yellow, blue at base; feet and legs, olive.

# APPENDIX

## KEYS TO THE FAMILIES, GENERA, AND SPECIES

## KEY TO THE FAMILIES.

| | |
|---|---|
| I. Toes of moderate length ; claws normal. | |
| *A.* Toes with more or less lateral membrane, sometimes scalloped. | |
| *a.* Bill slender and pointed, or flattened and blunt ; legs short, compressed ; sides flattened. | PHALAROPES. *Phalaropodidæ.* Page 21. |
| *B.* Toes without lateral membrane. | |
| *a.* Front of tarsus covered by small hexagonal scales ; hind toe almost always absent. | |
| *a'.* Bill straight or upturned at tip; legs not flattened, very long ; tarsus more than twice the length of hind toe and claw. | AVOCETS AND STILTS. *Recurvirostridæ.* Page 32. |
| *b'.* Bill slender, straight, basal part of culmen depressed, anterior portion arched ; always shorter than tarsus. | PLOVERS. *Charadriidæ.* Page 163. |
| *c'.* Bill longer than tarsus, much flattened, especially toward the tip, like a paper-cutter : basal web between outer and middle toes. | OYSTER CATCHERS. *Hæmatopodidæ.* Page 205. |
| *b.* Front of tarsus covered by transverse scales ; hind toe almost always present. | |
| *a'.* Bill slender, variable in length and shape, frequently flexible, sensitive, blunt or rounded at tip ; toes cleft to the base or partly webbed. | WOODCOCKS AND SNIPES. *Scolopacidæ,* Page 37. |
| *b'.* Bill rather stout, wedge-shaped, pointed at tip, about equal to tarsus, deep at base. | TURNSTONES. *Aphrizidæ.* Page 197. |
| II. Toes and claws excessively lengthened ; sharp spur on bend of wing. | JACANAS. *Jacanidæ.* Page 214. |

# THE PHALAROPES.

## FAMILY PHALAROPODIDÆ.

The birds comprised in this family are distinguished from all waders by having lateral membranes (usually scalloped) to the toes, similar to those possessed by the Grebes (*Colymbus*), and Coots (*Fulica*). There are only three species known, one of which is exclusively American. Apparently built for an aquatic existence, they possess with their lobed feet, a narrow, flat, compressed leg, and a thickened duck-like plumage to resist dampness and afford buoyancy and lightness to the body, and so render possible the quick, active movements which characterize the species when floating on the surface of the water. Three genera are here recognized, distinguished as follows:

### KEY TO GENERA.

*A*. Toes with broad-scalloped lateral membranes.

| | |
|---|---|
| *a*. Bill broad, flattened, stout; nostrils not near frontal feathers; central tail-feathers half an inch longer than outermost. | RED PHALAROPE. *Crymophilus*. |
| *b*. Bill slender, narrow, pointed; nostrils near frontal feathers; central tail feathers not half an inch longer than outermost. | NORTHERN PHALAROPE. *Phalaropus*. |
| *B*. Toes with narrow, even, lateral membranes, not scalloped. | WILSON'S PHALAROPE. *Steganopus*. |

There are either three recognizable genera in this family or only one genus. The Red Phalarope has been separated from the others on account of its broad, somewhat flattened bill, which is certainly very different in form from that of the Northern and Wilson's Phalaropes, which have a slender, delicate, and pointed bill. The lateral membrane of the toes, however, of the Red and Northern Phalaropes is similar, being wide and deeply scalloped at the joints, while that of Wilson's is narrow and without scallops. This seems to be as good a generic character as the bill, and if the Red is separated generically from the other two by the form of the latter, but agreeing with the Northern in the form of the lateral membrane, the Northern and Wilson's, though agreeing in the form of the bill, are equally separated generically by the shape of the lateral membrane of the toes. Regarding these characters as of equal generic importance, I have retained the species in three genera, restoring *Steganopus*, which has been suppressed in the A. O. U. check list. Stejneger, in his article on Phalaropus in the *Auk*, vol. ii., 1885, p. 183, does not refer to *Steganopus* at all, and makes no mention of the above differences between the Northern and Wilson's Phalaropes.

## *GENUS CRYMOPHILUS*

(Greek κρῦμόσ, *krumos*, ice cold, φίλεο, *phileo*, to love).

Crymophilus, Vieill, Analyse, 1816, p. 62. Type, *Tringa fulicaria*. Linn.

Bill broad, straight, flattened, nostrils near the base of maxilla, linear; tarsus, scutellated in front and behind; small hind toe present; marginal membrane of toes broad, scalloped at joints; web between outer and middle toe extending beyond second joint of latter.

One species only known, inhabiting both Old and New Worlds, and breeding in high latitudes.

## *GENUS PHALAROPUS*

(Greek φαλαρόπυς, *phalaropus*, coot-foot).

Phalaropus, Briss, Orn., vol. vi., 1760, p. 12. Type, *Tringa lobata.* Linn.

Bill slender, lengthened, pointed; nostrils basal; marginal web on toes as in Crymophilus.

One species only known, inhabiting both hemispheres in their northern portions, breeding in Arctic regions.

## *GENUS STEGANOPUS*

(Greek στεγανόπυσ, *steganopus*, web-foot).

Steganopus, Vieill, Nouv. Dict. Hist. Nat., vol. xxxii., 1819, p. 136. Type, *S. tricolor.* (Vieill.)

Bill similar to Phalaropus, but longer; web between outer and middle toe not reaching to second joint; lateral membrane narrow and barely scalloped.

One species known, restricted to America, largest of the family, of graceful and delicate form, with longer legs and better proportions than the others.

# THE AVOCETS AND STILTS.

## FAMILY RECURVIROSTRIDÆ.

THE two North American species of this family are, like their congeners in other parts of the world, rather large birds, remarkable chiefly for the shape of the bill, and the excessively long slender legs, which give to one of the species its trivial name, since it has the appearance of being elevated on a pair of stilts. One bird has the toes semi-palmated, and it is also noted for having the bill turned upward towards the point. The family is represented throughout the world by about a dozen species. Two genera are recognized as follows:

| | |
|---|---|
| *A*. Hind toe present. Anterior toes webbed. Bill recurved. | AVOCET. *Recurvirostra*. |
| *B*. Hind toe absent. Small web only between middle and outer toe. Bill straight. | STILT. *Himantopus*. |

### *GENUS RECURVIROSTRA*

(Latin *recurvus*, bent upward, *rostrum*, bill.)

Recurvirostra, Linn, Syst. Nat., vol. i., 1758, p. 151. Type, *R. avocetta*. Linn.

Bill long, slender, tapering, curved upward for a third of its length from the tip, which is slightly decurved. Hind toe present; anterior toes webbed. Legs very long; tarsus reticulated, covered with small hexagonal scales.

Four species of avocets are known scattered about the world, only one, however, being a native of North America. They are birds of singular appearance, with their curved-up bills and long legs, and are remarkable for their peculiar mode of feeding.

## GENUS HIMANTOPUS

(Greek *ἱμαντόπους*, himantopus, strap-leg.)

**Himantopus**, Briss, Orn., vol. v., 1760, p. 33. Type, *Charadrius himantopus*. Linn.

Bill straight, higher than broad, of moderate length. Hind toe absent. Outer and middle toe only connected by web. Legs very long, slender; thigh bare for a distance, and half as long as tarsus, which is nearly twice the length of middle toe.

There are about seven recognized species of Stilts, only one of which, however, is included in the North American fauna. It is chiefly noted for its excessively long legs, which seem hardly capable of upholding the body of the bird. It, however, walks well and runs easily, and when flying, which it does with firm even beats of the wings, the legs extend far beyond the tail.

# WOODCOCKS AND SNIPES.

## FAMILY SCOLOPACIDÆ.

This is the largest family of the waders, comprising about twenty species known to be natives of, or visitors to, North America. The vast majority of the birds seen running along our sea beaches, or probing the salt meadows and muddy tracts left bare by the tide, belong to this family. They are gregarious, associate and travel in large flocks, mostly migrants within the boundaries of the United States, and are noted for their graceful forms, attractive plumage of generally subdued colors, and gentle confiding dispositions, as well in some especial instances for the palatable quality of their flesh. To sportsmen they are the means of affording both with the dog or over decoys, most agreeable recreation with the gun, while a study of their habits yields much pleasure to many. Even the most inappreciative person can not regard without admiration the swift and graceful movements of many species of this family, as with easy flight they wheel and circle around him. They are the most attractive of all our dwellers by the sea.

The genera of this family here adopted amount to thirteen, some of which contain one or more sub genera, as *Tringa*, *Totanus*, etc. A number of the genera approach closely to each other, and it is not always easy to enumerate characters sufficiently trenchant by which those who are not ornithologists can always recognize one from the other, but I have endeavored in the table here given to diagnose these in such a manner as I trust will enable anyone, even if unfamiliar with a scientific method, to ascertain to what genus any bird included in this family, which he may obtain, shall belong. The general appearance of the members of the Scolopacidœ vary greatly both in whole or in part. The bill may be long, short, slender, stout, turned up at the end, or downward from the middle ; narrow at the point, or, as in one genus, expanded like a spoon. The legs also may be stout or slender, short or greatly lengthened, covered with broad scales behind and before, or with broad in front and small hexagonal ones behind. The toes too may be all partly webbed, or a web present at the base between two toes, or no web at all. Also the hind toe may be present, (as is usually the case), or absent entirely. And between these extremes are intermediate genera with allied characters, making the task of clearly defining them all difficult and perplexing.

KEY TO THE GENERA.

I. Ears placed beneath the eye. Plumage same at all seasons.

*A*. Bill straight, longer than tarsus and middle toe.

*a*. Markings on top of the head, transverse.

*a*¹. Under surface of tail feathers, with silvery white tips.

*a*². First three primaries longest, broad, of normal shape. } EUROPEAN WOODCOCK. *Scolopax*. Page 37.

| | |
|---|---|
| $b^2$. First three primaries shortest, attenuated, slightly curved. | AMERICAN WOODCOCK, *Philohela.* Page 39. |

*b*. Markings on top of the head, longitudinal.

| | |
|---|---|
| $a^1$. Under surface of tail feathers with pale buff tips. | SNIPE. *Gallinago.* Page 44. |

II. Ears placed behind the eye. Plumage changing with the seasons.

*A*. Bill straight, longer than tarsus and middle toe.

| | |
|---|---|
| *a*. Outer and middle toe connected by web at base. | DOWITCHERS. *Macrorhamphus.* Page 52. |

*B*. Bill straight or slightly curved, shorter than tarsus and middle toe.

*a*. Anterior toes webbed.

$a^1$. Back of tarsus with transverse scales.

$a^2$. Central tail feathers longest. Tail graduated.

| | |
|---|---|
| $a^3$. Tarsus twice as long as middle toe. | STILT SANDPIPER. *Micropalama.* Page 60. |
| $b^3$. Tarsus less than twice the length of middle toe. | SEMIPALMATED SANDPIPER. *Ereunetes.* Page 97. |
| $b^2$. Central tail feathers not lengthened; tail round. Tarsus half as long again as middle toe. | WILLETS. *Symphemia.* Page 129. |

*b.* Anterior toes cleft to the base.

*a*¹. Hind toe present.

*a*². Difference between length of the longest and shortest primary greater than length of bill. First primary longer than fourth. Nostrils not reaching beyond basal fourth of length of bill.

*a*³. Inner web of primaries uniform, not mottled. — SANDPIPERS. *Tringa.* Page 63.

*b*³. Inner web of primaries mottled. — BUFF-BREASTED SANDPIPER. *Tryngitis.* Page 144.

*b*¹. Hind toe absent. — SANDERLING. *Calidris.* Page 102.

*c.* Middle and outer toe united at base by a web.

*a.* Back of tarsus with scales like the front.

*a*¹. Frontal feathers extending beyond the base of gape.

*a*². Tail rounded about half the length of wing. Lateral groove on bill, extending about half the length. — TATTLERS. *Totanus.* Page 115.

*b*². Tail rounded, much less than half the length of wing. Lateral groove on bill extending nearly to the tip.

*a*¹. Bill over ⅓ length of wing; turned up toward tip. — GODWITS. *Limosa.* Page 105.

| | |
|---|---|
| $b^1$. Bill straight, less than 1-5 length of wing. | RUFF. *Pavoncella.* Page 136. |
| $c^2$. Tail, more than half the length of wing. | |
| $a^1$. Tail graduated. | UPLAND PLOVER. *Bartramia.* Page 141. |
| $b^1$. Tail rounded. | SPOTTED SANDPIPER. *Actitis.* Page 147. |
| *b*. Back of tarsus covered by hexagonal scales. | WANDERING TATTLER. *Heteractitis.* Page 134. |
| *C*. Bill very long, greatly curved from base. Toes cleft; back of tarsus covered with hexagonal scales. | CURLEWS. *Numenius.* Page 151. |
| *D*. Bill much widened and flattened at tip, its width at this part half the length of exposed culmen. Toes cleft. | SPOON-BILLED SANDPIPER. *Eurynorhynchus.* Page 95. |

## *GENUS SCOLOPAX*

(Latin *Scolopax*, a snipe).

Scolopax, Linn, Syst. Nat., vol. i., 1758, p. 145. Type, *S. rusticola.* Linn.

Body robust. Bill long, twice the length of tarsus. Primaries normal, first one longest. Tarsus scutellated both in front and behind.

This and the succeeding genus comprise the four known species of woodcock. Those of the genus Scolopax are recognizable by the broad primaries, from the American species. Only one, the European woodcock, represents the genus in North America, where it can only be considered as a straggler.

## *GENUS PHILOHELA*

(Greek *φίλοσ.*, *philos*, loving, *ἕλοσ*, *helos*, a bog).

Philohela, Gray, List Genera, 1841, p. 90. Type, *Scolopax minor.* Gmel.

The single species of the present genus can be distinguished from the other Woodcock of the genus Scolopax by the shape of the first three primaries, which are narrow and attenuated, curving slightly inwards. It is a smaller bird than its European relative, and of better flavor.

Only one species is known in North America, our familiar Timberdoodle, or Woodcock.

### *GENUS GALLINAGO*

(Latin *gallina*, a hen).

Gallinago, Leach, Syst. Cat. Brit. Mamm. & Birds, 1816, p. 31. Type, *Scolopax major*. Linn.

Bill long, slender, straight, depressed at tip, which is flexible. Bare part of thigh, scutellated before and behind, and reticulated on the sides like the tarsus. Middle toe longer than tarsus. Toes cleft to base. Tail of from twelve to twenty-six feathers; the North American species having usually sixteen. Plumage same in both winter and summer.

The species of this genus are about fifteen, scattered throughout the world, some being large, fine birds. Of the two species in the American fauna, one is indigenous to the country, the other is a straggler from Europe.

The Snipes, as they are called, as distinguished from others of their tribe, are shy, solitary birds, active at night, frequenting marshy ground, which they probe with their long bills in search of worms, etc., and have a swift, rapid, and erratic flight. They do not go in flocks like other waders, though at times many may be found congregated on one stretch of marshy ground.

#### KEY TO SPECIES.

*A*. Markings on head longitudinal. Tips of under surface of tail feathers, pale buff.

*a*. Tail with fourteen feathers. } ENGLISH SNIPE. *G. gallinago*.

*b*. Tail with sixteen feathers. } WILSON'S SNIPE. *G. delicata*.

### *GENUS MACRORHAMPHUS*

(Greek *μακρόσ*, *makros*, long, *ῥάμφοσ*, *rhamphos*, beak).

Macrorhamphus, Leach, Syst. Cat. Brit. Mamm. & Birds, 1816, p. 31. Type, *Scolopox griseus*. Gmel.

Similar to Gallinago, but at once distinguished by the web between the outer and middle toe at the base. Bill long, flattened and expanded at the tip. Tarsus longer than middle toe. Plumage in winter and summer, very different.

Two species comprise this genus, both indigenous to North America. They are similar in form to the species of Gallinago, but more slender, and with comparative longer bills and legs. The partly webbed foot at once differentiates the two genera. The habits are those of the waders, rather than of the snipes. They go in flocks, and inhabit marshy tracts near the sea, or in one species, the Long-billed Dowitcher, the banks of lagoons and rivers of the interior.

## KEY TO SPECIES.

*A*. Rump and upper tail-coverts, white, barred with black. Basal web between outer and middle toe.

*a*. Abdomen and under tail-coverts, white, latter barred with black. } DOWITCHER. *M. griseus*.

*b*. Entire under parts, cinnamon, under tail-coverts barred with black. } LONG-BILLED DOWITCHER. *M. scolopaceous*.

## *GENUS MICROPALAMA*

(Greek *μἰκρόό*, *mikros*, small, *παλάμη*, palame, a web).

Micropalama, Baird, B. N., Amer., 1858, p. 726. Type, *Tringa himantopus*. Bon.

Body slender; bill long, slender, straight, compressed, expanded at tip. Legs very long; tarsus twice as long as middle toe; thighs scutellated like the tarsus. Anterior toes all webbed at the base. Central tail feathers longer than the lateral ones. Wings long and pointed.

One species only is known belonging to this genus, a native of North America, migrating south in winter. More delicately formed than are the members of the preceding genus, it is yet closely allied to them, goes in flocks, frequents similar kinds of ground, and resembles the Dowitchers in its general habits. It is, however, not so often met with, and is altogether a rarer species, especially in its summer dress, individuals in that plumage being seldom seen.

## *GENUS TRINGA*

(Latin *Tringa*, a sandpiper).

Tringa, Linn, Syst. Nat., vol. i., 1758, p. 148. Type, *T. canutus*. Linn.

Bill straight, short, rather stout, widened near the tip, slightly longer than head. Tarsus short, nearly equal to middle toe and claw. Wings long and pointed, reaching beyond the tail in some species; difference between the length of the longest and shortest primary greater than length of bill. Inner web of primaries not mottled. Tarsus scutellated before and behind, and all species possess a hind toe. No webs between toes.

This genus is nearly a cosmopolitan one, containing a considerable number of species, about twelve of which are natives of, or visitors to, North America. They are dwellers of the sea coast and marshes near the ocean, are usually of small size, and have a powerful well-sustained flight, and run and walk swiftly and gracefully. Tringa has five subgenera, viz.: *Tringa*, *Arquatella*, *Actodromus*, *Pelidna* and *Ancylocheilus*; the first with tarsus and middle toe about equal, and bill straight; the second with tarsus shorter than middle toe and claw; the third and fourth with tarsus longer than the same, and the last with tarsus and middle toe about equal, and bill considerably curved. The key will enable any one to determine the species, which are there indicated in summer plumage.

## KEY TO THE SPECIES.

|  |  |
|---|---|
| *A*. Wing shorter than, or not reaching to, end of tail. | |
| *a*. Bill longer than tarsus, straight, or slightly curved downward. | |
| $a^1$. Bill straight, rather stout. | KNOT. *T. canutus*. |
| $b^1$. Bill slightly curved at tip, rather slender. | |
| $a^2$. No black on breast. | |
| $a^3$. Breast feathers in summer gray, margined with white. | PURPLE SANDPIPER. *T. maritima*. |
| $b^3$. Breast feathers in summer margined with buff. | COUES ALEUTIAN SANDPIPER. *T. m. couesi*. |
| $b^2$. Black on breast. | PRYBILOFF SANDPIPER. *T. m. ptilocnemis*. |
| *b*. Bill about equal to tarsus, straight. | |
| $a^1$. Lower neck and breast deep buff or rufous, legs and feet greenish yellow. | SHARP-TAILED SANDPIPER. *T. acuminata*. |

$b^1$. Lower neck and breast spotted or streaked. Legs and feet black.

$a^2$. Flanks brown, streaked with brownish black.

| | |
|---|---|
| $a^3$. Size large. Wing over 5½ inches. | COOPER'S SANDPIPER. *T. cooperi.* |
| $b^3$. Size smaller. Wing less than 5½ inches. | PECTORAL SANDPIPER. *T. maculata.* |
| $b^2$. Flanks white streaked with black. | WHITE-RUMPED SANDPIPER. *T. fuscicollis.* |
| $c^2$. Flanks pure white, sometimes tinged with buff. | BAIRD'S SANDPIPER. *T. bairdii.* |

*c*. Bill shorter than tarsus.

| | |
|---|---|
| $a^1$. Sides and front of neck buff, streaked with blackish. | LONG-TOED STINT. *T. damacensis.* |
| $b^1$. Sides and front of neck white, streaked with brownish black. | LEAST SANDPIPER. *T. minutilla.* |

*B*. Wings reaching beyond end of tail.

*a*. Bill much curved towards point.

$a^1$. Upper tail-coverts dusky. Belly black.

| | |
|---|---|
| $a^2$. Wing 4.30 to 4.75. Red on back dull. | DUNLIN. *T. alpina.* |
| $b^2$. Wing 4.60 to 4.95. Red on back bright. | RED-BACKED SANDPIPER. *T. a. pacifica.* |
| $b^1$. Upper tail-coverts white. Belly cinnamon rufous. | CURLEW SANDPIPER. *T. subarquata.* |

## *SUBGENUS ARQUATELLA.*

(Latin diminutive of *arcuata*, bowed).

*Arquatella* Baird, B. N. Am. 1858, p. 717. Type *Tringa maritima.* Brünn.

Form very compact or robust, the legs especially. Tarsus shorter than the middle toe with claw, the latter two-thirds to three-fourths as long as the bill, which is slender, much compressed, straight or very slightly decurved at the end.

In the A. O. U. check list three species are included under this subgenus, but it is very questionable if two of them, Coues' Sandpiper (*T. couesi*), and Prybiloff Sandpiper, (*T. ptilocnemys*), are entitled even to a sub-specific rank. Their alleged characters do not seem to hold good when a series of specimens are examined, and all that can be said of them is that *T. couesi* is lighter generally than *T. maritima* and *T. ptilocnemys* is lighter than either. Dimensions cannot be relied on, as the various parts differ greatly even in specimens from the same locality. *T. Ptilocnemys* was supposed to be confined to the Prybiloff Islands, in Behring Sea, but it is now known to inhabit a number of other islands, and to visit in winter the coast of Alaska, on the west, and the Kurile Islands on the east, and it is difficult to believe that it is a distinct species from its ally inhabiting similar areas. In the words of Seebohm, Charadriidæ, p. 431, "Such a geographical anomaly can only be accepted provisionally, pending further information." And as regards this form I would endorse the following statement of Baird, Brewer and Ridgway, B. N. Am. Water Birds, vol. 1, p. 223: "We therefore, all things considered, look upon the present bird (*T. m. ptilocnemys*), as being merely a local insular (?) race of a species of which *A. Couesi* represents the resident form of the coast of Alaska and the Aleutian chain, and from which *A. maritima* is perhaps not specifically distinct." For convenience I retain *Couesi* and *ptilocnemys* as subspecific forms.

## SUBGENUS ACTODROMUS.

(Greek ἀκτή, akte, seashore, δρομάς, dromas, running).

Actodromus, Kaup. Sk. Ent. Eur. Thierw. 1829, p. 37. Type *Tringa minuta*. Leisl.

Bill, slender, and little if any longer than the tarsus. The latter longer than middle toe and claw. Toes, slender, completely cleft. Wings, long, pointed.

## SUBGENUS PELIDNA.

(Greek πελιδνός, *pelidnos*, gray).

Pelidna Cuv. Règn. Anim. 1817, p. 490. 1829, p. 526. Type *Tringa alpina*. Linn.

Bill, slender, longer than tarsus, which is longer than middle toe. Bill decurved at the tip.

## SUBGENUS ANCYLOCHILUS.

(Greek, ἀγκυλόχειλοϛ, *agkulocheilus*, curved bill).

Ancylochilus, Kaup. Ent. Eur. Thierw., 1829, p. 50. Type *Tringa ferruginea*, Brünn.

Bill, similar to Pelidna, but much more curved.

## GENUS EURYNORHYNCHUS

(Greek εὐρύνω, *euruno*, I dilate, ῥύγχοϛ, *rhugchos*, beak).

Eurynorhynchus, Nilss, Orn. Suec., vol. ii., 1821, p. 29. Type, *Platalea pygmæa*. Linn.

Bill spatulate, spoon-shaped, three times as wide near the tip as it is at the base. Toes not webbed. Size of body small, rather short, rounded.

One singular species composes this genus, the curious Spoon-bill Sandpiper, accidental on our western shores, a straggler from Asia. It is at once recognizable from all the waders by its singularly formed bill, that spreads greatly towards the tip, and gives that member the shape from which it derives its trivial name.

## GENUS EREUNETES

(Greek ἐρευνητήϛ, *ereunetes*, a prober).

Ereunetes, Illiger, Prodromus, 1811, p. 262. Type,' *Tringa pusilla*. Linn.

Anterior toes webbed at the base; hind toe present. Bill slightly expanded at tip, about as long as tarsus. Size small.

Two species are known, among the smallest of the Sandpipers, inhabiting North America. Some authors include the species of Macrorhamphus and Macropalama in the present genus, but there is no doubt that sufficient characters exist to separate generically the birds arranged under these three divisions.

The Peeps are among the most numerous of the Sandpipers, as well as the smallest, and familiar to every one who has tramped over the salt marshes and mud flats adjacent to the sea.

Their habits and flight are like those of the other members of the tribe. The species may be defined as follows:

Size, small. Feathers of back margined with buff or cinnamon. Front toes webbed at base.

| | |
|---|---|
| Culmen, .68 to .92 inches. | SEMIPALMATED SANDPIPER. *E. pusillus.* |
| Culmen, .85 to 1.15 inches. | WESTERN SANDPIPER. *E. occidentalis.* |

## GENUS CALIDRIS

(Greek καλίδρισ, *calidris*, name of some beach bird).

Calidris, Cuv. Leç. Anat. Comp., vol. v., 1805, Tabl. ii. Type, *Tringa arenaria.* Linn.

Characters similar to those of Tringa, but the hind toe is wanting. Bill straight, slightly broadened at the tip. Middle toe barely two-thirds the length of tarsus. No web on the foot.

Only one species is known, distributed generally throughout the world, and easily distinguished from all other Sandpipers by having no hind toe, in this respect resembling the Plovers.

## GENUS LIMOSA.

(Limus mud, mire).

Limosa, Brisson, Orn. vol. v. 1760, p. 261. Type, *Scolopax limosa.* Linn.

Bill, long, grooved, curving slightly upward towards the tip, which is not attenuated, and exceeding the tarsus in length, and over one-third length of wing, with lateral groove extending nearly to the tip. Tarsus, transversely scutellated behind and before, reticulated laterally. Middle and outer toes, united by a membrane at the base, extending to the first joint on the outer, basal on the inner toe.

The Godwits are rather large birds, remarkable for their long bills, which have a slight upward tendency towards the tip. This part of the maxilla is somewhat thickened and extends beyond the end of the mandible. They are shore birds, but are often found far inland, where they frequently breed. The genus contains but few species, four only being recognized as belonging to the North American Fauna. The following key will designate the species :

| | |
|---|---|
| *A.* Rump and upper tail-coverts buff, diagonally barred with dark brown. | MARBLED GODWIT. *L. fedoa.* |
| *B.* Rump dark brown, bordered with white; upper tail-coverts white, barred with dark brown. | PACIFIC GODWIT. *L. l. baueri.* |
| *C.* Rump black. Upper tail-coverts white. | |
| *a.* Axillaries brownish black. | HUDSONIAN GODWIT. *L. hæmastica.* |
| *b.* Axillaries white. | BLACK-TAILED GODWIT. *L. limosa* |

## *GENUS TOTANUS.*

(Italian *totano*, name of a bird).

Totanus, Bechst, Orn. Tasch. Deutschl, 1803, p. 282. Type *Scolapax totanus.* Linn.

Bill, slender, straight, or slightly inclined upward at the tip, shorter than tarsus, lateral groove of maxilla extending over basal half. Toes, with small webs at base. Tarsus, half as long as middle toe, but twice as long for subgenus *Glottis*, and nearly equal in length for subgenus *Rhyacophilus*. The back covered with scales like the front,

This genus is represented throughout the world. Its members frequent seacoasts as well as inland pools and banks of rivers, breeding generally inland and in Arctic or Antarctic regions. They walk gracefully, run swiftly, and have a firm, well sustained and rapid flight. They swim easily, and some can even dive. The note is a clear, musical whistle, and they are usually very gentle and confiding. The nest is a depression in the ground. It has two subgenera, *Glottis* and *Rhyacophilus*, the former including the so-called Yellow-legs or Tattlers, and the latter the Solitary Sandpipers; the slight characters distinguishing them are given above.

The following table will enable the various species recognized in North America to be easily distinguished:

### KEY TO THE SPECIES.

*A.* Upper tail-coverts white, barred with black.

*a.* Rump brownish black, feathers edged with grayish white. Legs yellow.

| | |
|---|---|
| $a^1$. Bill over two inches in length. | GREATER YELLOW-LEGS. *T. melanoleucus.* |
| $b^1$. Bill less than two inches, about one and a half. | LITTLE YELLOW-LEGS. *T. flavipes.* |
| *b*. Rump white. Legs and feet greenish yellow. | GREENSHANK. *T. littoreus.* |
| *B*. Middle upper tail-coverts blackish brown, spotted on margin with white. | |
| *a*. Spots on upper parts and wings white. | SOLITARY SANDPIPER. *T. solitarius.* |
| *b*. Spots on upper parts and wings brownish cinnamon. | WESTERN SOLITARY SANDPIPER. *T. s. cinnamomeus.* |
| *C*. Upper tail-coverts pure white. | GREEN SANDPIPER. *T. ochropus.* |

## *SUBGENUS GLOTTIS*

(Greek γλῶττα, *glotta*, the tongue).

Glottis, Koch, Baier, Zool., 1816, p. 304. Type, *Totanus glottis*. Bechst.

Middle toe half as long as tarsus.

## *SUBGENUS RHYACOPHILUS*

(Greek ῥύαξ, *hruax*, a brook, φίλοσ, *philos*, loving).

Rhyacophilus, Kaup. Sk. Ent. Eur. Thierw., 1839, p. 140. Type *Tringa glareola*. Linn.

Middle toe as long as tarsus.

## *GENUS SYMPHEMIA*

(Greek σύμφημι, *sumphemi*, I speak with).

Symphemia, Raffinesque, Jour. de Phys., vol. vii., 1819, p. 418. Type, *Scolopax semipalmata*. Gmel.

Bill stout, straight, strong, compressed laterally. Feathers falling short of the nostrils, going a little farther forward on mandible than on maxilla. Bill about equal to tarsus, the latter half as long again as middle toe. Base of all toes webbed. Wings long, pointed. Tail rounded.

One species of this genus only is known, with one subspecies, if it is really sufficiently distinguishable to entitle it to such recognition. The Willet is among the largest of this family excepting the Godwits and some of the Curlews. It is distributed in its two forms over the whole of temperate North America, and is a conspicuous and well known member of the waders. The genus is not represented in the Old World.

| | |
|---|---|
| *A.* Upper parts grayish brown, blotched with black. Under tail-coverts white, barred with dark brown. Culmen, 2¼ inches. | WILLET. *S. semipalmata.* |
| *B.* Upper parts grayish drab, faintly marked with black. Under tail-coverts pure white or faintly barred with brown. Culmen, 2½ inches. | WESTERN WILLET. *S. speculifera.* |

## *GENUS HETERACTITIS*

(Greek ἕτερος, *heteros*, different, actitis, a genus of snipes).

Heteractitis, Stejneger, Auk, 1884, p. 236. Type, *Scolopax incanus.* Gmel.

This genus, signifying a difference from certain Shore Birds—viz., of the genus Actitis, contains but two species, only one of which at present is included in our fauna. The characters of the genus are: Bill straight, rather stout, longer than tarsus, with the nasal groove extending over the basal two-thirds. Tarsus covered laterally and behind with hexagonal scales, and tibia likewise covered with similar scales, differing in this respect from the other species of *Totaneæ.* Outer and middle toe connected by web, reaching to first joint of the latter; and middle and inner toes by a rudimentary web. Hind toe nearly one-third the length of tarsus. They are rather solitary birds, inhabiting rocky shores.

## *GENUS PAVONCELLA.*

(Diminutive of *Pavo*, peacock).

Pavoncella, Leach, Syst. Cat. Brit. Mamm. & B., 1816, p. 29. Type *Tringa pugnax.* Linn.

Bill, straight, tapering, point flattened, higher than broad at base, one-fifth length of wing; nasal groove extending nearly to the end; nostrils basal, linear; first primary, longest; legs, long, slender; thigh, bare for one-third its length; tarsus, one and a quarter as long as middle toe, scutellate, outer and middle toes connected at base by a web. Face of male in summer covered with fleshy tubercles, and neck covered by a ruff of lengthened feathers.

But one species of this genus is known, a native of the Old World, where it has a wide distribution; stragglers occasionally visit North America, records of which are sufficiently numerous to cause the adoption of the species into our fauna. It is remarkable for the great diversity of color and pattern of markings in the plumage of the male in breeding dress, and for various curious habits.

## *GENUS BARTRAMIA.*

(In honor of Wm. Bartram).

Bartramia, Less. Trait. Ornith. 1831, p. 553. Type, *Tringa longicauda*. Bechst.

Bill, long, slender, nearly as long as head, straight, grooved nearly to tip; nostrils, near the base, linear; wings, long, pointed; tail, long, graduated, more than half the length of wing; legs, long, slender; tibia bare for fully a third of its length; tarsus, half as long again as middle toe with claw. Web between outer and middle toes from the first joint of the former, merely at base between middle and inner toes.

But one species is known of this genus, essentially belonging to the New World, with occasional stragglers to the eastern hemisphere, as far as Australia. It is a bird of the uplands, not frequenting the sea-coast, nor salt marshes, and is a graceful, prettily plumaged species.

## *GENUS TRYNGITES.*

(Greek, τρύγγαϭ, triggas or tringas, sandpiper),

Tryngites. Cab. Jour für Ornith. 1856, p. 418. Type, *Tringa subruficollis*. Vieill.

Bill, straight, maxilla grooved to near the tip, shorter than head. Feathers extend to nostrils on maxilla, beyond on the sides of the mandible, and cover the inter-ramal space. Wings, long, pointed. Inner web of primaries mottled; tail, long, doubly emarginate; legs, long, slender; tarsus, a little longer than the middle toe and claw. Toes cleft to the base.

One species represents this genus, resembling in its habits the previous one, Bartram's Sandpiper. It is peculiar to the New World, straggling frequently to Europe. Small in size, tame and confiding in disposition, and clothed in plumage of attractive hues, it is a most pleasing object wherever found.

## *GENUS ACTITIS*

(Greek ἀκτή, *akte*, seashore).

Actitis, Illiger, Prodromus, 1811, p. 262. Type, *Tringa hypoleucus*. Linn.

Bill longer than the head, straight, rather slender, both maxilla and mandible grooved. Wings, long pointed. Legs of moderate length; toes long; outer connected with middle by a large membrane; inner slightly connected. Tail, more than half the length of wing, rounded.

Two species are included in this genus, one a native of the New World, occasional in Europe; the other found in northern parts of Old World and some islands in Behring Sea, but never obtained in our limits. They frequent the banks of brooks and rivers, lay their eggs on the sand or on the pebbly beach, and are remarkable for the tilting movement of the hinder portion of the body, which gives them a ludicrous aspect. They are gentle and unsuspicious, and the American species is one of the most familiar objects along the lakes and rivers.

## GENUS NUMENIUS

(Greek νέοσ, *neos*, new, μήνη, *mene*, the moon, crescent-shaped bill).

Numenius, Brisson, Ornith., vol. vi., 1760, p. 311. Type, *Scolopax arquata*. Linn.

Bill very long, but variable in length among individuals of the same species, always longer than the tarsus, culmen rounded and curved downward for half its length from the tip; maxilla longer than mandible. Tip of bill expanded, and the grooves reaching to the middle. Feathers of chin reaching forward to anterior end of nostrils. Legs moderately long; lower half in front covered with scutellated plates, behind by hexagonal reticulations, distinguishing them from Sandpipers and Snipes. Outer toe attached to middle toe by a web extending to first joint; web to inner toe of about half the size. Hind toe small. Wings long, first primary longest, tertials nearly as long as primaries. Tail short, even.

There are nearly a dozen species belonging to this genus distributed over all the earth, three only of which are strictly natives of the New World, and two others occasional visitors to North America. They are large birds, remarkable for the lengthened decurved bill, which makes them conspicuous objects in any group of waders of which they may form a part, or on the marshy land or moor over which they walk with stately steps. As a rule they are shy and wary birds, taking flight at the slightest alarm with shrill loud cries.

The following key will distinguish the species:

*A.* Bill long, greatly curved downward,
maxilla longer than mandible,

| | |
|---|---|
| *a*. Axillaries rich dark buff without bars. | LONG-BILLED CURLEW. *N. longirostris.* |
| *b*. Axillaries dark buff, barred with dark brown. | |
| $a^1$. Shafts of thigh feathers not prolonged beyond web. | |
| $a^2$. Top of head blackish brown, with vertical buff stripe. Tail rufous, barred with dark brown. | HUDSONIAN CURLEW. *N. hudsonicus.* |
| $b^2$. Top of head streaked with black and buff. Tail grayish brown barred with dark brown. | ESKIMO CURLEW. *N. borealis.* |
| $b^1$. Shafts of thigh feathers prolonged beyond the webs. | BRISTLE-THIGHED CURLEW. *N. tahitiensis.* |
| *c*. Axillaries white, barred with blackish. Rump white. | WHIMBREL. *N. phæopus.* |

# THE PLOVERS.

## FAMILY CHARADRIIDÆ.

THIS family includes the true Plovers, so-called, (those species with usually three toes on the foot, the hind one absent), and their allies. It is a very large group, represented throughout the world, and its composition has been a subject of frequent discussion among ornithologists as to what species and genera should be placed in it. Happily there is not much doubt regarding the American species, although one genus, Aphriza, appears to be an aberrant form, and of somewhat distant relationship, and is properly placed in a different Family. About fifteen species are recognized as North American, two or three of which, however, can only be regarded as stragglers within our limits from the Old World. They are mostly gregarious, some going in immense flocks, frequenting various portions of the Continent, from both sea-coasts, inland. They run and fly with swiftness, and have a soft, melodious note, according well with their gentle dispositions. Keys are given for the genera, of which there are three recognized, and of the species under each genus, when it contains more than one.

### KEY TO GENERA.

*A*. Head crested. Plumage on upper parts metallic. } LAPWINGS. *Vanellus.*

*B*. Head not crested.

| | |
|---|---|
| *a*. Plumage on upper parts spotted in black, white, or golden. | GOLDEN PLOVERS. *Charadrius*. |
| *b*. Plumage on upper surface unspotted, uniform. | RING PLOVERS. *Ægialitis*. |

## GENUS VANELLUS.

(Diminutive of *vannus*, a fan).

Vanellus, Brisson, Orn. vol. v. 1760, p. 94. Type *Tringa vanellus*. Linn.

Bill shorter than the head, about equal to the middle toe in length, straight, and decurved at tip, which is blunt. Nostrils situated at base of maxilla in a groove that extends two-thirds the length. Outer toe connected to the middle toe by a web at base. Hind toe with claw present. Head with a long occipital crest. Wings, long and rounded, the first primary about equal to the sixth. Legs, long, slender, tarsus scutellated in front, recticulated on side. Tail, moderately long, nearly even. Plumage metallic.

There are about four species belonging to this genus, all of which are natives of the Old World, one only having been occasionally procured in North America, Greenland, and Long Island, on the east, and Alaska on the west. It is a very handsome bird, with a metallic plumage, and the head ornamented with a graceful recurved crest.

## GENUS CHARADRIUS.

(Greek χαραδριὸό, *charadrios*, a plover).

Charadrius, Linn, Syst. Nat., vol. i., 1758, p. 150. Type, *C. apricarius*. Linn.

Legs reticulated with five or six rows of hexagonal scales in front, fewer behind. Hind toe rudimentary or obsolete. First primary longest. Tail rounded.

Four species of this genus are natives of, or visitors to, different portions of North America, and these have been placed in two subgeneric divisions according to the presence or absence of a hind toe. The genus is cosmopolitan, found in the northern portions of both hemispheres. They frequent open grassy places, most generally found on the uplands, are gregarious, sometimes going in very large flocks.

| | |
|---|---|
| *A*. Hind toe rudimentary. Axillaries black. | BLACK-BELLIED PLOVER. *C. squatarola*. |
| *B*. Hind toe absent. | |
| *a*. Axillaries white. | EUROPEAN GOLDEN PLOVER. *C. apricarius*. |
| *b*. Axillaries smoky grey. | |
| *a*[1]. Wing averaging in length 7.09 inches (6.80–7.20). | AMERICAN GOLDEN PLOVER. *C. dominicus*. |
| *b*[1]. Wing averaging in length 6.40 inches (6.10–6.80). | PACIFIC GOLDEN PLOVER. *C. d. fulvus*. |

## SUBGENUS SQUATAROLA

(Italian name of the species).

Squatarola, Cuv. Règn. Anim., vol. i., 1817, p. 467. Type, *Tringa squatarola*, Linn.

Hind toe rudimentary.

## GENUS ÆGIALITIS.

(*αἰγιαλίτηό*, a dweller by the sea).

Ægialitis, Boie, Isis., 1822, p. 558. Type, *Charadrius hiaticula*. Linn.

Bill small, compressed, as high as broad, about as long as middle toe without claw. Nostrils placed in a skin situated in a groove, which extends beyond centre of bill. Wings long, pointed, reaching nearly to end of tail; first primary longest; inner secondaries nearly as long as primaries. Tarsus nearly twice the length of middle toe, covered with hexagonal scales; no hind toe. Anterior toes slender, slightly webbed at base.

This genus includes the Ring Plovers, so called on account of the breast having one or more bands of different colors crossing it more or less completely. It contains about a dozen species, the majority small in size, dwellers of the seashore for the most part, quick runners and rapid flyers, and possessed of clear, soft, melodious voices.

The American group has four subgenera and ten species, one at least of which is very doubtfully included in our fauna. The following key distinguishes these.

## *SUBGENUS OXYECHUS.*

(Greek ὄχυσ, *oxus*, sharp, ἦχοσ, *hechus*, sound).

Oxyechus, Reich, Syst. Av., 1853, p. xviii. Type, *Charadrius vociferus.* Linn.

Tail very broad, long, two-thirds length of wing.

## *SUBGENUS ÆGIALITIS.*

See Genus.

Size small. Tail half as long as wing. Type, *Æ. semipalmata*

## *SUBGENUS OCHTHODROMUS.*

(Greek ὄχθοσ, *ochthos*, bank, δρομάσ, *dromas*, running).

Ochthodromus, Reich, Syst. Av., 1852, pl. xviii. Type, *Charadrius wilsonius.* Ord.

Bill, long, stout, longer than middle toe. End of culmen curved. Basal half of maxilla lower than terminal half.

## *SUBGENUS PODASOCYS.*

(Greek πόδασ, *podas*, ὠκύσ, *okus*, swift as to his feet).

Podasocys, Coues, Proc. Acad. Nat. Scien. Phil., 1866, p. 96. Type, *Charadrius montanus.* Towns.

Bill slender, wide at base, longer than middle toe, which is less than half the length of tarsus.

## KEY TO THE SPECIES.

*A.* Black band below the white on back of neck.

*a.* Two black bands across the breast. Rump ochraceous. } KILLDEER. *Æ. vocifera.*

*b.* One black band across the breast.

$a^1$. Back and rump ashy brown.

$a^2$. Base of both maxilla and mandible, yellow.

|  |  |
|---|---|
| $a^3$. White spot behind eye nearly obsolete. Web at base of toes to second joint. | SEMIPALMATED PLOVER. *Œ. semipalmata.* |
| $b^3$. White spot behind eye conspicuous. Web on toes only to first joint. | RING PLOVER. *Œ. hiaticula.* |
| $b^2$. Base of mandible only yellow. | LITTLE RING PLOVER. *Œ. dubia.* |
| $b^1$. Back and rump whitish ash, or brownish gray. | |
| $a^2$. Black band on breast not meeting in center. | PIPING PLOVER. *Œ. meloda.* |
| $b^2$. Black band on breast continuous. | BELTED PIPING PLOVER. *Œ. m. circumcincta.* |
| *B*. No black band on back of neck. | |
| *a*. Crown and nape reddish buff. Sides of breast black. | SNOWY PLOVER. *Œ. nivosa.* |
| *b*. Crown and nape brownish gray, uniform. | |
| $a^1$. Bill moderate. Band across breast cinnamon rufous. | MONGOLIAN PLOVER. *Œ. mongola.* |
| $b^1$. Bill large. Band across breast black. | WILSON'S PLOVER. *Œ. wilsonia.* |
| *C*. Crown and entire upper parts grayish brown, feathers edged with rufous, or reddish buff. Across breast an indistinct ochraceous band. | MOUNTAIN PLOVER. *Œ. montana.* |

# THE TURNSTONES.

## Family Aphrizidæ.

This Family includes the two known species of Turnstones, one of which is a cosmopolitan bird, found throughout the world, together with the rather widely distributed but little known species generally called "Surf Bird" of the genus Aphriza, from which the Family takes its name. They are essentially "Beach-birds," frequenting rocky shores and sandy stretches along the borders of the ocean, and have the habit of turning over small stones wherever these abound, to search for such insects or small crustacea as lay beneath and on which they feed. They are very rapid runners, swift flyers, breed in the vicinity of the sea, and utter a clear, loud whistle. One species, the Black Turnstone, is indigenous to North America.

### KEY TO THE GENERA.

*A*. Bill at base higher than broad, pointed.

*a*. Bill plover-like, middle part depressed. Culmen shorter than tarsus. } APHRIZA.

*b*. Bill compressed, culmen straight, about equal to tarsus. } ARENARIA.

## Sub-Family Aphrizinæ.

### *GENUS APHRIZA.*

(Greek, ἀφρόσ, *aphros*, sea-foam, ζάω, zao, I live).

Aphriza. Aud. Orn. Biog. vol. v. 1839, p. 249. Type *Tringa virgata*. Gmel.

Bill, Plover-like, middle portion depressed, terminal portion swollen and arched Culmen shorter than tarsus. Tail emarginate. Toes cleft, hind toe prominent.

But one species is known, inhabiting the Pacific coast of America. In some respects it resembles the Plovers, but has a very prominent hind toe, and presents many points of resemblance to the Turnstones.

It has been associated with the Plovers by some writers, but it is very obvious that its place is apart from that group, and much more naturally associated with the species of the following genus.

## SUB-FAMILY ARENARIINÆ.

### *GENUS ARENARIA.*

(*Arenaria* relating to sand; a sand pit).

Arenaria, Briss. Orn. vol. v. 1760, p. 132. Type *Tringa interpres.* Linn.

Bill, compressed, pointed at tip, culmen straight. Culmen and tarsus, nearly equal. Tail rounded. Hind toe present.

Two species only are known belonging to this genus, one of them cosmopolitan, the other restricted to North America. They are birds of moderate size, one with very gay and attractive plumage of strongly contrasting colors of black, dark red and white. They are exclusively shore birds, nesting near the beach.

### KEY TO THE SPECIES.

*A.* Rump, and under parts below breast, pure white.

| | |
|---|---|
| *a.* Back and scapulars, blotched with rufous or light chestnut. | TURNSTONE. *A. interpres.* |
| *b.* No red or chestnut on any part of plumage. | BLACK TURNSTONE. *A. melanocephala.* |

# THE OYSTER CATCHERS.

## FAMILY HŒMATOPODIDÆ.

The Oyster Catchers are very peculiar looking birds, rather large of size, having a plumage, with two exceptions, of strongly contrasted colors, and with bright colored legs and bill, the latter of unusual shape. They are birds of more temperate climes, in which all breed, but some also nest in Arctic as well as tropical regions. Nearly cosmopolitan in their distribution, and dwellers of the seashore and island beaches, they are shy and wary, keeping well away from possible danger. The bill of the Oyster Catcher is of peculiar shape, and distinguishes these birds from all other groups of waders. About seven species are recognized, scattered over various parts of the world.

### *GENUS HŒMATOPUS*

(αιμα, *haima*, blood, πούς, *pous*, foot, red-footed).

Hœmatopus, Linn, Syst. Nat., vol. i., 1758, p. 152. Type, *Hœmatopus ostralegus*. Linn.

Bill, flat, compressed, wedge-shaped, tapering to a point, longer than the tarsus. Nasal groove extending beyond centre of maxilla. Tarsus covered by hexagonal scales; no hind toe; a basal web between outer and middle toe. Legs and feet stout. Wings long, pointed. Tail short, nearly even.

The species of this genus are among the largest of the waders, stout and heavy in body, wary, and have a swift and powerful flight. They live and breed upon the coasts near the sea, run with speed, and walk, it may almost be called "stalk," along the shore with some approach to a dignified carriage. Four species are included in the North American fauna, one of which, a European form, is occasionally found in Greenland, one is generally distributed in our limits, and two are restricted to the Pacific Coast.

## KEY TO SPECIES.

| | |
|---|---|
| *A.* Legs and feet dull crimson. | EUROPEAN OYSTER CATCHER. *H. ostralegus.* |
| *B.* Legs and feet pale flesh color. | |
| *a.* Upper tail-coverts pure white. | AMERICAN OYSTER CATCHER. *H. palliatus.* |
| *b.* Upper tail-coverts varied with brown and white. | FRAZAR'S OYSTER CATCHER. *H. frazari.* |
| *c.* Entire plumage blackish brown. | BLACK OYSTER CATCHER. *H. backmani.* |

# THE JACANAS.

## FAMILY JACANIDÆ.

THE Jacanas are peculiar birds, formed for a partially aquatic existence and provided with excessively lengthened toes, which enable them to move with ease and safety over the broad leaves of water plants lying on the surface. The claws also are long, that on the hind toe excessively lengthened, much longer than the toe itself. The Family is divided into five genera, containing nine species, which are pretty well distributed throughout the globe. The wing is armed at the bend with a spur, which in a number of species is moderately long and very sharp, in others blunt.

### *GENUS JACANA*

(*Jacana*, native name for the bird).

Tail short, central feathers not lengthened beyond the rest. Primaries of normal shape. Head with an upright fleshy wattle on top. Toes and claws excessively lengthened.

Four species comprise this genus, natives of Mexico, Central America and Northern South America to Brazil. One species strays into Texas, and so becomes a member of the North American fauna. They are pretty birds, rather pugnacious in disposition, and the spur with which Nature has endowed them apparently for purposes of offence and defence, is a formidable weapon. These birds are striking objects in the lily-covered ponds and lagoons scattered about the localities they frequent.

# INDEX.

www.ingramcontent.com/pod-product-compliance
Lightning Source LLC
Chambersburg PA
CBHW021350210326
41599CB00011B/826